Perspective on Social Forestry

By the same author

Dryland Farming Perspectives and Prospects

Perspective on Agro Ecological Problems : A Geographical Study in Environmental Impact Assessment.

Nature of Geography as Perceived in Systems Theory in 2 Vols. : Vol. 1 : Systems Theory and Geography, Vol. 2 : Possible World of Geographer in Systems Theory

PERSPECTIVE ON SOCIAL FORESTRY

B.L. Sharma
M.A.; Ph.D.

&

R.L. Vishnoi
M.A.; Ph.D.

2018

Daya Publishing House®

A Division of

Astral International Pvt. Ltd.

New Delhi - 110 002

First Published, 2000
Reprinted, 2018

ISBN: 978-81-7035-228-0 (Hardbound)

Publisher's note:

Every possible effort has been made to ensure that the information contained in this book is accurate at the time of going to press, and the publisher and author cannot accept responsibility for any errors or omissions, however caused. No responsibility for loss or damage occasioned to any person acting, or refraining from action, as a result of the material in this publication can be accepted by the editor, the publisher or the author. The Publisher is not associated with any product or vendor mentioned in the book. The contents of this work are intended to further general scientific research, understanding and discussion only. Readers should consult with a specialist where appropriate.

Every effort has been made to trace the owners of copyright material used in this book, if any. The author and the publisher will be grateful for any omission brought to their notice for acknowledgement in the future editions of the book.

Published by : **Daya Publishing House**®
A Division of
Astral International Pvt. Ltd.
– ISO 9001:2008 Certified Company –
4736/23, Ansari Road, Darya Ganj
New Delhi-110 002
Ph. 011-43549197, 23278134
E-mail: info@astralint.com
Website: www.astralint.com

Acknowledgements

We owe gratitude of thanks to all those officials and personnel, who have directly or indirectly co-operated and provided informations and data for the present work. It is our earnest duty to express thanks to Mr. B.S. Rathore and Mr. M.L. Ameta for their help in preparing this work.

B.L. Sharma

R.L. Vishnoi

Contents

List of Tables

List of Figures

List of Charts

List of Abbreviations

1. Av.	Average.
2. C. S. S.	Centrally Sponsored Schemes.
3. D. F. O.	District Forest Office.
4. Dy. C. F.	Deputy Conservator of Forests.
5. F. D.	Forest Department.
6. F. W. P.	Fuel Wood Plantation.
7. ha.	Hectare.
8. J. F. M.	Joint Forest Management.
9. L. P. G.	Liquified Petroleum Gas.
10. N. F. P.	National Forest Policy.
11. N. G. O.	Non Governmental Organisation.
12. N-N-Road	Nashirabad-Neemuch Road.
13. N. S. F. P.	National Social Forestry Project.
14. PAN.	Protected Area Network.
15. P. F. M.	Participatory Forest Management.
16. P. F.	Protected Forests.
17. P. S.	Panchayat Samiti
18. R. B. H.	Rehabilitation of Barren Hills.
19. R. D. F.	Rehabilitation of Degraded Forests.
20. R. F.	Reserved Forests.
21. Rmt.	Running Meter.
22. R. O.	Range Officer.

23. S. F.	Social Forestry.
24. S. P. P.	Silvi-pastoral Plantation.
25. T. P.	Timber Plantation.
26. U. F.	Unclassed Forests.
27. V.F.P. & M. C.	Village Forest Protection and Management Committee.
28. V. F. W.	Village Fuel Wood.

Chapter-1

Introduction

"Trees means water, water means bread and the bread is life"

(K.M. Munshi, 1950)

Importance

Consciously or unconsciously, the land has been the major source of livelihood of the people. For this reason, continued misuse of land over a long period has been subjected to its use in future.

In agricultural systems, forestry is considered a vital link between human efforts and the ecology of the area. It provides fuel, fodder and fruits as well as protects from the climatic vagaries. Grazing and fodder are virtually free from the forests all over the country. But in a few cases, as has been a common practice in the rural side, house wife has to spend a day or two, to gather fuelwood for a week or so. This has an undesirable impact on the home life and particularly the care of the children, in addition to the continuous destruction of forests in neighbourhood. This phenomenon makes the birth of 'Social Forestry' or forestry for people.

Thus, in the words of Shah, S.A. (1988), the 'Social Forestry' - (SF), may be described as "any programme of tree planting by any organisation on any land, including farm forestry is a programme of Social Forestry to the extent it meets the needs of the rural people".

Fuel, fodder and fruits (F^3) are the basic needs of the people, dwelling in rural side. It is the economy of the people on which they sustain life and go ahead. In beginning, population was less and resources were abundant, hence there was no danger of environmental disorder or degradation. But with the gradual span of time, the population increased with an alarming rate and disturbed the man-land ratio, resulted in widespread destruction of forests and degradation of ecosystem.

This problem stress the need of the preservation and conservation of forest wealth, and restoration of ecosystem, as an immediate measures of protecting human lives from any other natural calamity and disasters.

Social forestry in real sense, is the outcome of the forth-coming danger and increasing environmental distress all over the world. Thus, it is an old science of forestry, with a new terminology. It is said to be as old as 125 years (Shah, 1988) with its new phrase of "Social Forestry and People". Thus it is the people's participation in forestry system. The increasing distress of environmental degradation, land, and people and wide destruction of forests in the recent past has drawn attention of the foresters and scientists. The changing ecosystem, declining production and increasing wastelands are the glaring incidences of the environmental distress. Down cutting of forest trees for fuel needs, has declined rainfall, with increasing temperature.

Despite the resistance by the people in the beginning the importance of the social forestry was realised at the later stage.

The state forest department has very rightly observed the situation and stated that at one time large portions of the state were covered by trees, but their destruction for various purposes has very greatly reduced the area under tree crops and now only 3 per cent of the total land area of the state is under effective tree cover. The rest 6 per cent is in the shape of barren hills or degraded forests. The present fuelwood requirement in the state is approximately 90 lakh tonnes, were as the production of fuelwood from the government owned forests is about only 3 lakh tonnes. Therefore, the deficiency in forest resources is acute. The productivity of 60 per cent of the forest area is very low partly because of aridity an low soil fertility and partly due to the adverse biotic factors both human and bovine. The fuelwood situation in the state is very serious rather critical. Someone has rightly made an observation that we will have enough food but will not have enough fuelwood to cook it. In recognisation of above fact the task of forest management is now to find ways of adding new objective of meeting the local needs, the needs of rural communities.

In the present context of the problem as stated above, the importance may be drawn, as the state has very meagre source of livelihood on one side and limited forests cover on the other. Thus the district of Bhilwara is not an exception to this fact that it has very limited resource to meet the requirements of the people, with particular reference to the fuel and fodder.

It is in this context that the concept of forestry is lately gaining ground. Establishment of tree growth in areas nearer to the habitation aiming at meeting the local needs of fuel, fodder, food, fibre, flosses, small timber etc. for rural communities, is the central theme of social forestry. Such activities cover a wide variety of land ownership including Government. Waste lands, community pasture lands, village common lands, farmers fields, farm boundaries, homesteads, road/railway or canal sides and even marginal lands where agriculture has limited potential.

Aims and Objectives

Looking to the above problem the present study is focussed, mainly on to the people's participation in forest generated products. It is already clear from the above statement that forest management schemes are aimed to meet, people's requirements *i.e.* fuel and fodder. Thus the following schemes are implemented with the fulfilment of different objectives under social forestry;

(*i*) Agro forestry,

(*ii*) Fuel wood plantation,

(*iii*) Silvipastoral plantation,

(*iv*) Timber wood plantation,

(*v*) Rehabilitation of degraded forests (R.D.F.) etc.,

Under these objectives, it is understood that the Panchayat or the community land schemes in different ways fulfil the local needs of the people. For the purpose the community land is separated from the rehabilitation of degraded schemes. Rehabilitation schemes are generally employed in degraded forest areas.

Future Scope

The study entails with the fulfilment of fuel and fodder and to provide the future scope, to the people. As it has been mentioned above that social forestry is a people's participation, meaning by to engage rural people in different activities. These activities are related to the cultivation of plants, fuelwood and fodder for the animals.

In Bhilwara district, as it is discussed that the availability of fuel and fodder is estimated as 357.3 thousand tonnes of fuel and 2617.5 thousand tonnes of fodder or 62.5 and 78.4 per cent of fuel and fodder respectively. While the actual need of fuel and fodder is listed as 571.6 and 3339.4 thousand tonnes respectively. Thus there is shortage of 37.5 and 21.6 per cent of fuel and fodder against the actual need, which has to be raised from the social forestry schemes.

So far the achievements of the required fuel and fodder, through the number of plants and area covered under them, are recorded as 48.9 lacs and 5446.5 hectare land respectively.

It is also a future target that near about 117.5 thousand hectares land is expected to bring under the plantation schemes.

Coverage and Methodology

The book envisage with the aims and objectives, which includes the wide range of social forestry activities. These activities are generally covered with the farm-forestry and extention forestry. The farm-forestry includes the strip plantation and block plantations, where as the extention forestry is further added with village wood lots, aforestation and agro-forestry.

Thus it is very clearly brought out the wide range of areas covered under the social forestry schemes. The breeding ground of these schemes is the community land, locally known as Panchayat land. By selecting each of the community land in eleven Panchayat Samities, in the district, the case studies have been done to find out the actual achievements. These case studies, some way or the other provides, the man and environmental relationship and their eco-system for the future development.

On the basis of the empirical studies (case studies) the analytical explanations of each of the community land plantation have been done.

It is in the light of analytical explanations the impact assessment has been drawn from the environmental point of view. The social forestry and the environmental relationship in different parts is a key to the all developmental schemes.

Chapter-2

Land Use Systems in Bhilwara

General

Land in general and agricultural land in particular as a major resource, plays an important role in country's development. Agricultural land as a multifunctional single set of occupation for the largest section of the people has a special bearing as a principal resource. Because it is the Agriculture land which plays a crucial role in economic development, by way of food production. Thus it is a matter that how we make the use of the agricultural land.

Land as a resource being used for since long by the people for various purposes. But agricultural land is being put for specific use right from the dawn of human civilization. In this respect the agricultural land put to various purposes is a major concern to all of us.

Agricultural Land Use

The term 'land use' refers to the general categories of land used for various purposes. Whereas the agricultural land use refers to the land used for cultivation purposes. The term agriculture has a wider meaning, that includes the cultivation of crops, live-stock breading, fishing, forestry and poultry farming etc. In this context the agriculture land use refers to the land used for production purposes, which has a spatial dimention.

This be taken as land put to use diffently for different purposes (production) according to the need of farmers as best be selected for the purpose of earning some thing from the farms.

Thus agricultural scenerio becomes a varied with colourful cropping patterns as land uses, whether forestry plantation, livestock grazing or any other activity. According to Technical Committee on Coordination of Agricultural Statistics (T.C.C.A.S.),[1] the agricultural land has been classified into nine major categories for the agricultural land use purposes (Table 2.1).

Table 2.1 : Land use in Bhilwara District (1992 - 93)

S. No.	*Land categories*	*Area in 000 ha.*	*Percentage to reported area*	*Percentage to respective categories*
1.	Geographical area	1045.50	—	—
2.	Reported area for land utilization purpose	1045.50	—	—
3.	Forests	69.80	6.68	—
4.	Land not available for cultivation	219.00	20.95	—
(*i*)	Non agricultural land	58.00	5.55	26.48
(*ii*)	Barren and uncultivated land	161.00	15.40	73.52
5.	Other uncultivated land	293.10	28.03	—
(*i*)	Permanent pastures and grazing land	119.82	11.46	40.88
(*ii*)	Land under miscellaneous tree crops	0.38	0.04	(0.12)
(*iii*)	Cultivable waste	172.90	16.53	59.00
6.	Fallow land	102.60	9.81	—
(*i*)	Long fallow	59.30	5.67	57.79
(*ii*)	Current fallow	43.30	4.14	42.21
7.	Net area sown	361.00	34.53	—
8.	Area sown more than once	145.00	13.87	—
9.	Total cropped area	506.00	48.40	—

Source : District Statistical abstract - Bhilwara 1994. Directorate of Economics and Statistics Rajasthan, Jaipur.

In this respect the agricultural land use categories in Bhilwara district may be explained and analysed in the following forms.

In the district the total geographical area or the reporting area is recorded as 1045.5 thousand hectares, of which forest cover is found on 69.8 thousand hectare or 6.68 per cent as per revenue records. The land not available for cultivation is occupying 219 thousand hectares which comes to 20.95 per cent. This category of land is classified into two sub-categories *i.e.* (i) Non-agricultural land, occupying 58 thousand hectare (5.55 per cent) and (ii) Barren and uncultivated land occupying 161 thousand hectares (15.4 per cent). In this category it is interesting that barren and un-cultivated land is occupying about 73.5 per cent land, of the land not available for cultivation.

The land under other cultivated land includes the permanent pastures, land under tree crops and cultivable waste about 293.10 thousand Hectares (28.03 per cent) of the total geographical land, of which permanent pastures occupying 119.8 thousand (11.46 per cent), land under miscellaneous tree crops is 0.38 thousand hectares, cultivable waste land occupying 172.9 thousand or 16.53 per cent area of the total geographical land. It is interesting to note that cultivable waste land occupying 59.0 per cent area of this category and 40.88 per cent falls under permanent pastures. This reflects that the scope for cultivation was a wide range, if facilities are made available.

In the category of fallow land occupies 102.6 thousand hectares which is about 9.8 per cent of the geographical area. Under this category long fallow occupying 5.67 per cent and current fallow 4.14 per cent area of the district. It is to note that 57.79 per cent land has a long fallow, that shows the inadequacy of the resource management.

In the overall geographical and the net area sown (NAS) accounted for about 34.5 per cent or 361 thousand hectares. This reflects the land under cultivation is very much less as it has much scope of agricultural land, if the long fallow and agricultural waste land is brought under cultivation, in the district. The total cropped area is reported as 506 thousand hectares which accounted for 48.4 per cent of the geographical area, shows that about 13.9 per cent area is the area sown more than once or double cropped. Thus nearly 50 per cent area of the total geographical area comes under cultivation.

Agricultural Land Use Systems

Agricultural land use systems are the true reflection of the age-old practices carried on in different parts of the world. In fact the agricultural land is used in different ways under cultivation. In doing so one has to take care that soil nutrients may be retained and carrying capacity of land is restorted. This depends on the various methods of cultivation in which soil efficiency, fertility and productivity is maintained. For that purpose the following land systems ar adopted.

(*i*) Fallow system,

(*ii*) Agro-Forestry System,

(*iii*) Agro-Silvipastoral System, and

(*iv*) Dry-land Farming System.

(i) Fallow System

In agricultural practices fallow system is an important of the agricultural land use systems. It is common practice to leave land (field) or a portion of land without a crop for a particular period of time. In order to regain the vitality of the soil. There are two types of fallow systems *viz.;* current fallow, and long fallow system, in former case land is left without any crop at least for one or two years. While in the later case, land is left for more than two years.

Crops like cotton. Maize, millets etc. extracts more soil nutrients and hence, soils are exhausted for the next cropping. Therefore, such land is left vacant for a period to regain vitality. Sometimes land is left vacant for a longer period, so as to soils may recover lost fertility.

In the district context, it is reported that 102.6 thousand hectares or 9.81 per cent of the total geographical area is put under fallow systems. The long fallow is accounted for about 5.67 per cent where as current fallow 4.14 per cent. This type of land is accounted for 28.42 per cent of the NAS.

(ii) Agro-Forestry System

Agro-Forestry is a long practised system. In every agrarian society there is a practice of alternate resource mobilization. In agricultural land use systems land is not extensively put to farming practices, because of some natural and Socio-economic constraints. But for alternate source of income farmers go for forestry in order to obtain fuel, fodder, fruits, fibre and flosses. Besides these farmers get many other things like wood, charcoal and other forest products also.

This system is very much useful in dry climate of Rajasthan where dryness is a common feature. The cultivation of tree along with fields is a climatic necessity to protect crops.

(iii) Agro-Silvi-Pastoral System

This system is almost related to the Agro forestry system but with a little change of activity related to the live-stock ranching. Some times fodder grasses are cultivated side by side along with cropping under treeshade. In such farming system farmers developing alternate source of lively-hood by way of live-stock breeding, tree-collections and cultivation of crops. Therefore, these activities are supplementing each other, specially when there is natural hazards.

(iv) Dry Land Farming System

It is a practice of cultivation of crops without irrigation. Actually in areas of inadequate water, some crops are cultivated. In this sense dry land farming is a 'rainfed agriculture'[2]. It refers to the cultivation of crops from the soil moisture;[3] and also called 'water concentrating culture'[4]. In river basin watershed areas are delineated on which some crops are grown just after kharif season crops like pulses, oil seeds, pearl millets and fodder crops are cultivated specially in such watershed area; better known as dry land farming. In dry areas of Rajasthan crops like Baira, Jaowar, Maize, Pulses and oil seeds are occupying wider areas of the cultivated land. It constitute about 70 per cent of the cultivated area.

Field-Systems

Unlike agricultural land use systems there are some field-systems, used to maintain land potentiality. These field-systems invariably used by the farmers in one way or the other to obtain more production. But in fact it is a practice of maintaining soil productivity. Field systems are restricted only to the operational land areas *i.e.* land put to crops. In fact field systems are important from the Agro-ecological point of view. Because these practices provide conditions of bacterial growth. Some times they are useful for the development of micro-organs, but at times they also develop harmful bacterias to sustain upon favourable conditions.

In general the following field-systems are adopted as;

(*i*) Crop rotation system

(*ii*) In-field and out-field system

(*iii*) Mixed cropping system

(*iv*) Inter-cropping system, and

(*v*) Alley cropping system.

(i) Crop Rotation System

In this system the cultivated land is rotated according to the suitability of crops. This practice is very useful in regaining the vitality of the soil. For example after harvesting cotton, groundnut is grown, as cotton exhausts soil nutrients while groundnut provides the lost nutrients. Thus suitable combination of crops is adopted invariably.

The following combination of crop rotations are adopted in cultivation of some crops :

(*a*) Bajra by moth, mung and oil seeds.

(*b*) Mustard by groundnut/pulses.

(*c*) Jaowar by pulses, gram/til.

(*d*) Wheat/barley by bajra/maize.

The above mentioned crop-rotation is adopted in accordance with the environmental factors, which help in maintaining eco-system. Therefore, the practice of field system becomes a necessary part of the Agro-ecology in a region. A best crop rotation helps in restoring the balance between soil structure and the environmental factors.

(ii) In-field and Out-field Systems

This system is very common in the traditional bound agrarian society. Actually it is a system of restricted field practices. The in-field and the out side field system, both are practiced simultaneously by farmer on his field when he uses operational practices, *e.g.* while irrigating fields, some times he may leave half of the portion of the field without irrigation. Similarly when he applies manure to the field he may have to abandon a portion of the field. In doing so, the fields getting irrigation or manure is called as the in-field, where as the field or field portions left without water or manure is called out-field. In other words the fields which are within the reach of the farmer and out of range are called in-field and out-field. Therefore these practices mainly depending on the resources available to the farmers, in time.

(iii) Mixed Cropping

This system, actually is a part of subsistence type of agriculture, where farmers cultivate more than one crop simultaneously. Some times they intermix crops and sometimes they are grown on a portion of the field.

(iv) Inter Cropping

Actually this system has been developed from the mixed cropping. In this system two or three crops are cultivated in a field, as the farmers possess small size of holdings. Thus it forms a subsistence type of agriculture. Farmers cultivate two or three crops simultaneously in suitable conditions with a suitable combinations. The best suited crops are mung with Bajra or Gaur with Bajra, Til with Moth, Maize with Urd and Chanwala. Fodder crops alongwith other grains form an important part in this system.

(v) Alley-Cropping

Alley cropping system is adopted in areas of constant soil erosion. In this system the cultivation of crops is done along with the tree-rows aligned with the contours. It is an important system in almost all the agrarian traditional bond societies to maintain the agro-ecology. While doing so farmers are benefitted in two ways *viz.* protecting crop form the sun and chilly winds and maintaining ecology. These trees in turn provide wood as well as tree products to the farmers. The important species of such trees are subabool, Sisbania, Khejari, Ardu, Caliandra spps. These trees provide forage cum mulch, gum etc. ...

From the Social Forestry point of view alley cropping system forms an important part in the present study, because it is the alley cropping system which helps in maintaining the balance between the man and environment, in other words, the eco-system.

Social System in Agriculture

Farming is an occupation as well as a way of life, reflects the responses of the individual's performance under a given set of environment. In any society the social attributes play an important role in shaping economic development. Societies are known after their ways and means which they have acquired inheritance. It is true reflection of their cultural ways. Thus, their behaviour acts in a particular manner to develop economic means. Agriculture in a wider sense is an occupation as well as cultural way of living of the people and a major source of livelihood. Thus it reflects the culture, social ways and the economic practices on the whole.

In this respect every society in general and agrarian society in particular, with their cultural and social bondage acts in a particular manner to perform agricultural activities. Therefore the social systems are important part of the agrarian social structure. It is the social structure which gives rise to a particular type of agriculture. The agricultural systems in this sense are the true reflection of the social systems. In traditional agrarian societies the following principal social systems are at work.

(*i*) Land Ownership System,

(*ii*) Land Tenural System, and

(*iii*) Size of Land Holding

(i) Land Ownership System

Land ownership "refers to the agriculture land or a part of it owned or held in owner like possession as an individual, joint or corporate (Company), private property". Thus land possessed or supervised by an individual or joint as a legal owner or Khatedar is called land ownership.

Thus the land ownership which is considered as an important institutional frame work affects the pattern of land utilization, because the legal frame work determines the right of ownership and the distribution of land.

In the district context land ownership equally plays an important role as discussed in the above pages. In the district it has two facets of the land ownership *viz.* individual and joint. In the district there are about 175 thousand individual and 101 thousand joint ownerships.

(ii) Land Tenural System

The concept of land tenure is taken here to mean the system or individual agreement whether written or not, under which land is held or occupied[5]. The concept of tenancy system is therefore "refers to the agricultural land that is operated under the labour (in exchange of services) or share tenancy or any other form of landed bondage"[6]. In this respect the land is operated under certain conditions laid down by the owner of the land.

Land tenancy is an age old system prevailing in the traditional agrarian societies. The nature of this system is as such that in large number peasants do not possess land, therefore they had to work on other's field but with conditions. In this system the major share of production goes to the owner of the field, therefore it is nothing but a simple exploitation of poor farmers (marginal) by the hands of land lord or Jamindar.

There are three major forms of tenancy system.

(*a*) Share of cropping,

(*b*) Share of Produce, and

(*c*) Lease or contract basis.

(a) Share of Cropping

This system is an oldest one in which maximum benefit goes to the owner. In this system the cultivated crops on a field is shared by the owner and the tenant. While doing so it it agreed that 1/4th or so part of field goes to the tenant while rest is shared by the owner. Generally the good portion of the cropping is shared by the owner. Thus the peasant has to suffer the losses of crop damage.

(b) Share of produce

It is a bit developed form of share of cropping system in which instead of crop, the produce is shared by the owner and tenant. Thus the tenant in return gets at least something of the agreed part.

In most of the countries this type of system is generally expected by the larger section of the agricultural labourers who are working as tenants.

(c) Lease or Contract Basis

This type of tenancy is a further development of share of produce. Because the peasants gradually realised that in both the first forms, they have to suffer at the hands of land owners. Therefore they thought it better to possess cultivated land on a contract basis or on a lease to operate upon them. Though, it is not a common practice for all farmers but in some areas those who can afford practicing on these lines may take up. Farmers usually take cultivated land on 2 or 5 years contract or even more than this and intensively cultivate the fields. The owner gets the agreed money.

The social system at work in the agrarian societies as an institutional way makes an important point in the social forestry when it becomes a part of Agro-ecology. Because in disturbing the Agro-ecological balance man and his institutional frame-work comes to an end. Thus the present study makes a point to have a wider perspective of the field operation and its associated form of social ways.

(iii) Size of Land Holding

In the social set up of agrarian society is the land which counts much more. The farmer's status is judged from the land he possesses but on the other side it also refers to the size of land he owns. Thus both the factors, the ownership and size of land make an important issue while dealing with the agricultural operations.

In traditional society like ours, the social practices are considered vital for the agricultural development. Whether it may be a state, district or a village the social set up and the practices are the common factors for agricultural development. Thus land ownership or tenancy system or the size of land holdings are the true reflections of the society which are equally operative in either of the forms.

Thus land holding system is very effectively working in all the traditional bond societies. Land holding has two forms :

(*a*) Pieces,

(*b*) Fragmentation.

The term "Pieces" denotes that the holding is cut into pieces according to the number of sons in a farmer's family (head). By this practice the total size of holding is reduced to pieces. This practice is a common in all the traditional countries, where son by a social bondage (legal) automatically becomes the owner of the property (by inheritance) after the death of the head of the family.

Where as the fragmentation refers to the holding size which scattered in a different parts of the village (area). It is a typical feature of the social ways that a farmer posses a big size of holding (10 Hectare or more) but in different parts. This means the fields are found scatteredly in different parts.

If the farmers have a big holding as a cluster in a place then it makes a point of good agriculture. In a fragmented form farmer has to divide his labour, time and energy, thus the returns are not expressed as economic.

In the district the size of holding in different size groups are reported as shown in Table 2.2.

From the Table 2.2 it may be explained that there are about 175 thousand individual holdings and 101 thousand joint holdings in the district. The area covered by them is about 318.5 and 262.9 thousand hectares respectively.

It is interesting to note that holdings in the size group of 1.0 to 2.0 ha. are recorded highest number of holdings *i.e.* 29.46 per cent while in the size group of 0.5 to 1.0 ha. 22.42 per cent as second highest and 0.2 to 0.5 ha; 18.57 per cent as the third highest. This shows that about 70.45 per cent of the total holdings are reported under less than 2.0 hectares. According the National Sample Survey (NSS) this category comes under the marginal holdings. Therefore it may be said that about 70 per cent of the total operational holdings are recorded under 2.0 hectares.

Land holdings of 2.0 to 3.0 hectares size group records 13.6 per cent, 3.0 to 4.0 - 6.6 per cent and 4.0 to 5.0 - 3.4 per cent. This shows that in medium size group (2.0 to 4.0 ha.) includes 20.2 per cent of the total holdings. Under the size group of 10 hect. And above accounted for about hardly 1 per cent of the total holdings.

So far the area under different size group of the holdings is concerned we find that about 34.8 per cent area is concentrated, whereas in 2.0 to 4.0 hectares it constitutes 30.4 per cent and in the size group of above 10 hectares the area falls about 8.7 per cent. Thus it may be said that the small size of holdings covering more

Table 2.2 : Number of Holdings in Different size group : District Bhilwara

Size of holding	Individual holding Number	Percentage	Area (ha.)	Percentage	Joint holding Number	Area (ha.)
Below - 0.02	133	0.07	1.69	Ng	365	4.64
0.02 - 0.50	32496	18.57	10093.54	3.17	12632	4144.73
0.50 - 1.0	39221	22.42	28421.90	8.92	15897	11556.86
1.0 - 2.0	51543	29.46	72341.97	22.71	23538	34151.45
2.0 - 3.0	23753	13.57	57364.58	18.00	15451	37772.04
3.0 - 4.0	11552	6.60	39579.97	12.42	9758	33850.84
4.0 - 5.0	5961	3.41	26493.03	8.32	6282	28030.62
5.0 - 7.50	6277	3.59	37427.05	11.75	7833	47536.46
7.50 - 10.0	2243	1.29	19036.23	5.98	3201	27173.86
10.0 - 20.0	1574	0.89	20399.32	6.40	2319	29885.79
20.0 and above	226	0.13	7419.17	2.33	316	8821.52
Total	174979	100.00	318578.45	100.00	101192	262928.81

Source : Agricultural Statistical Report 1991 - District Bhilwara.

area giving rise to a subsistence type of agriculture. There is a little scope for commercialised and specialised agriculture, as the holdings are not very big. So far the impact on social forestry is concerned, the size of holdings reflects that the subsistence type of agriculture not very much favouring the forestry culture.

Waste Land

The concept of waste land is wider one in the context that scientists have not come to the common agreement about it's exact defination. Usually this term is applied to the land which is lying vacant and not being used for the specific purpose. In geography the waste land is that which is not being used for cultivation for some reasons, but may be brought under cultivation if resources are made available.

In the district this type of land is accounted for about 172.9 thousand hectares which comes to 16.6 per cent of the total geographical area of the district. In the category of other cultivated land but not being used, accounted for about 58.5 per cent which is sufficiently a very high ratio. Thus it may be said that in the district besides all other land which may be termed as the waste lands, the cultivated waste land makes a point.

Community Land

The community land in general may not be described in a particular sense except, that of land under trible community and religious or owned by a particular caste group. In the district this type of land is generally categorised as religious or *'Dev-Sthan'*. It has been a social practice to provide land for the temple (Math) etc. This type of land is very restricted and limited in size, therefore need not to emphasize in detail.

Here in this context it makes an point, that the community land possesses by the 'Gram Panchayat' in a specific sense is considered important from the social forestry point of view. Because it is the rightful duty of the 'Gram Panchayat' to develop such land under forestry cover and allied activities for the benefit of the village people. In other words this act is useful from the Social Forestry and environmental point of view.

References

1. Singh, Jasbir. 1974-75, An Agricultural Atlas of India. Kurukshetra University of Haryana.
2. Sharma, B. L. 1991, Dry land Farming. Daya Publications, New Delhi.
3. Falkner, F.R. 1938, The Desert Limit of Dry Farming in Africa. Geographer-84.
4. Jatzold, R. 1979, Rainfed Water Concentrating Culture, Applied Geography and Development. Vol. 21, Berlin.
5. Chauhan, B. S. 1966, Studies in Land Utilization.
6. Sharma, B.L. 1983, Agricultural Typology of Rajasthan, Udaipur (Raj.).

Chapter-3

Problem of Social Forestry

"A growing tree is a living symbol of a progressive nation"

—Jawahar Lal Nehru

Tree is a single unit of a forest. Whereas tree species are considered a family unit of a forest which constitute a community. The aggregate of tree communities (species) is a forest.

The Forest has a specific meaning. According to the glossary of technical terms[1] forest in general is "An area set aside for the production of timber and other forest produce, or maintained under woody vegetation for certain indirect benefits, which it provides *e.g.*, climatic or protective". Ecologically "A plant community predominantly of trees and other woody vegetation usually with a closed canopy".

Whereas legally "Forest is an area of land proclaimed to be a forest under a forest law. "Community forest", A forest owned and generally managed by a community such as village, town, tribal authority or local Unit, the members of which share in the produce or producers.

Whatever may be the forms or kinds of species for the welfare of the society and the people, it is the vegetative cover which makes an important point of the present problem. The term forests has a wider complications, when we go into the details of their classification. But for the present context in general usages the forests have been classified into three major categories *viz.*

(*i*) Reserved forest (R. F.),

(*ii*) Protected forest (P. F.), and

(*iii*) Unclassed forest (U. F.).

But, this classification is generally used for the administrative purpose. It reflects that the forest conservation is in otherwise form an essential part of the forestry system. The very idea of the forestry system coincides with the exploitation of forests on one side and the conservation of forests on the other.

The history reveals that the human civilization at the beginning flourished alongwith the river banks as well as in forest sides. In span of time those forest covers were removed in order to install better human civilization. Thus destruction of forests alongwith conservation makes an important point in the study of social forestry.

Forest conservation in India is about 125 years old. The foundations were laid and traditions established by German and British foresters. Some of the best preserved forests of the world are still to be found in India.

In spite of such strong well established traditions, it is unfortunate that forests have been subject to unimaginable pressure and consequent destruction in recent years. The increase in human and cattle population virtually has wiped out the tree growth from the official forest boundaries which one time served as a buffer for the forests. About 90 per cent of the wood produced in the country is used as fuelwood by about 80 per cent of rural population. People have to cook food and the cattle have to serve fodder any way. Encroachments and shifting cultivation have aggravated the situation under such circumstances. One sees scant prospects for preserving whatever is left out of forests, if the present situation continues unabated.

Concept of Social Forestry

The concept of social forestry was enunciated in a comprehensive manner for the first time by Mahatma Gandhi, the father of the nation. Pragmatic and down to earth as he was, he stated in clear terms without coming any new phrases[2].

"An ideal village should be so conditioned that the villagers could procure all daily needs and requirements of materials of thatch, bamboo, fuelwood, grazing etc. from within a radius of five miles"[3].

Looking to the above statement one may come to a complex matter of social forestry. Social forestry should be viewed not as a rigidly targeted tree planting programme to be achieved in a particular time frame, but as a complex process of interaction with farmers all geared towards improved standards of rural living. It is argued that to be meaningful to up land farmers, the emphasis in a social forestry programme should shift from trees to protecting and sustaining soil productivity. Trees are just used as one of the vehicles in achieving these goals. Solutions for successfully integrating trees into cropping systems should be found, taking the crops as the key link and recognizing that the farmers perceptions are oriented towards benefits that accrue from farming.

In fact the household unit is a critical factor in implementation of social forestry. 'The households' changing structure and composition brought about by its own development, provide the need and the opportunity for labour, necessary for the implementation of social forestry programme'[4].

Definition

Being a new subject in modern terminology, it is very difficult to define social forestry. Scientists, Agronomists and even environmentalists have not come to the common consensus for the exact definition of social forestry. Therefore a better definition of SF is yet to be foreseen. Credit may be given to National Commission on Agriculture (NCA) to coin the term "Social Forestry" K. M. Tiwari has tried to explain the term social forestry that what is 'being practised as social forestry is not in any way different from the conventional forestry, being practised well over a century'[5]. For coining the term 'social forestry' for the first time goes to Westoby[6] (1968), who defined it as "forestry which aims at producing flow of production and recreation benefits for the community". This definition has no resemblance whatsoever to that given by NCA which invested a broad base to it.

The main thrust of the National Commission on Agriculture in this programme was to develop tree resources largely outside the recognised forest boundaries in the neighbourhood of rural areas, irrespective of weather the land earmarked for producing such resources for community, private or government. Thus, conceptually, it has a broad base. A lot of criticism has generated in the academic as well as professional circles as to the precise meaning of social forestry and its relevance. In a specific terminology it may be defined as 'programme of tree planting by any organisation on any land, including farm forestry is a programme of social forestry to the extent it meets the need of rural people'[7].

Traditionally, forest-base domestic needs are used to be met from the community lands and other common property resources. As a result of continued use and misuse over a long period, without any effort of reforesting such lands, they become highly degraded and degenerated. Consequently, it is the economically backward section of the society, which depended on them relatively more than any other section, was the worst affected.

This point was clearly recognised by the National Commission of Agriculture, which developed the concept and philosophy of social forestry as is being implemented in the country to day. It is, therefore, very clear that, conceptually, social forestry deals with (Physically) sick lands and (economically) sick people, to produce goods such as fuel, fodder, etc., in order to meet the needs of the local community, particularly the under privileged section. Needless to say that, the National Commission on Agriculture derived its inspiration from Mahatma Gandhi as quoted earlier.

Aims and Objectives of Social Forestry

Forests were managed largely for the production of industrial wood. There was hardly any conscious managemental design for meeting exclusively or even substantially, the day to day needs of the people. Under the pressure of growing population, the demand for timber, fuelwood and bamboo increased substantially. Indian agriculture and rural way of life are highly dependent on forests and forest products (Chart-3.1). In short, the rural economy leans heavily on trees. It is this dependence which prompted the late Shri K.M. Munshi to coin a very meaning and appropriate phrase". "Tree means water, water means bread and the bread is life". Nothing could illustrate this dependence more than the amount of forest commodities utilized largely by the rural people.

In rural areas both agricultural waste and animal dung have alternate uses, which are more important nationally than using merely as fuel. However, in the absence of availability of fuel wood the rural poor are driven to use them. They are conscious of the fact that cowdung is used as a fuel instead of a valuable manure, is a national waste. They are faced with the choice; to use cowdung or to eat raw food. It is relevant to point out here that for the rural poor, fuel is largely non-monetised. They are too poor to purchase it.

On account of the reasons indicated above the National Commission on Agriculture looks ahead in the scope of Social Forestry in the following ways :

A. Farm Forestry

(*a*) Raising rows of trees on bunds or boundaries of the fields and individual tree in private agriculture lands.

(*b*) Wind break.

B. Extension - Forestry

(*a*) Mixed forestry, raising of grasses and leaf fodder together with fruit and fuelwood trees on suitable waste lands of Panchayat and village commons.

(*b*) Shelter Belts.

(*c*) Raising of plantations of different quick-growing species on side of roads, canal banks and railway lines.

(*d*) Reafforestation in degraded forests.

(*e*) Recreation forestry.

By the very nature of farm forestry, long rotations are out of question. It produces, small size poles suitable for rural housing. Such poles are in short supply and hence costly. Farm forestry under the circumstances, serves, a social purpose.

Farmers are motivated more by money than anything else. It is a question of relative economies between two alternate uses of land. *viz.* agricultural and farm forestry.

If the land is fertile enough to sustain a high degree of agricultural production, it is certain, he would not divert such lands to tree farming. It is the poorest of the fertile lands that is being diverted to social forestry. Because of a high unit value of timber and fuelwood. Agriculture has encroached upon forest lands a since long.

Thus in 1950 'Van Mahostava"[8] an annual festival of raising and protecting trees started celebrating with the laid down principles of social forestry with the following objectives to :

(*i*) provide fuel and thus to release cowdung for use as manure,

(*ii*) increase production of fruits, and thus add to the potential food resources of the country,

(*iii*) help, create shelter-belts around agricultural fields to increase their productivity,

(*iv*) help conservation of soil and stop further deterioration of soil fertility,

(v) provide leaf fodder for cattle and thus to relieve intensity of grazing over reserved forests (R. F.),

(*vi*) provide shade and ornamental trees for landscape,

(*vii*) provide small poles and timber for agricultural implements, house construction and fencing,

(*viii*) inculcate tree consciousness and love for trees amongst the people, and

(*ix*) popularize the planting and tending of trees in farms, villages, municipal and public lands for their aesthetic, economic and protective values.

In parallel to the above objectives Rowe. V. Cadelina[9] in a ford foundation Report SURADPU first surveyed project 1983, has suggested some objectives of the social forestry, which may be considered a little extension of the above mentioned objectives.

(*i*) To increase production level of individual household farms,

(*ii*) To protect the remaining forests from further cutting,

(*iii*) To restore the watershed functions of the deforested areas by introducing effective agro-forestry cropping system,

(*iv*) To sustain farm productivity,

(*v*) To achieve social justice by assisting dislocated populations to find their own niches, and

(*vi*) To assess inter ethnic competition in using locally available resource bases, production changes, soil conditions (availability and fertility), pest incidence and control, and the regeneration of permanent vegetation.

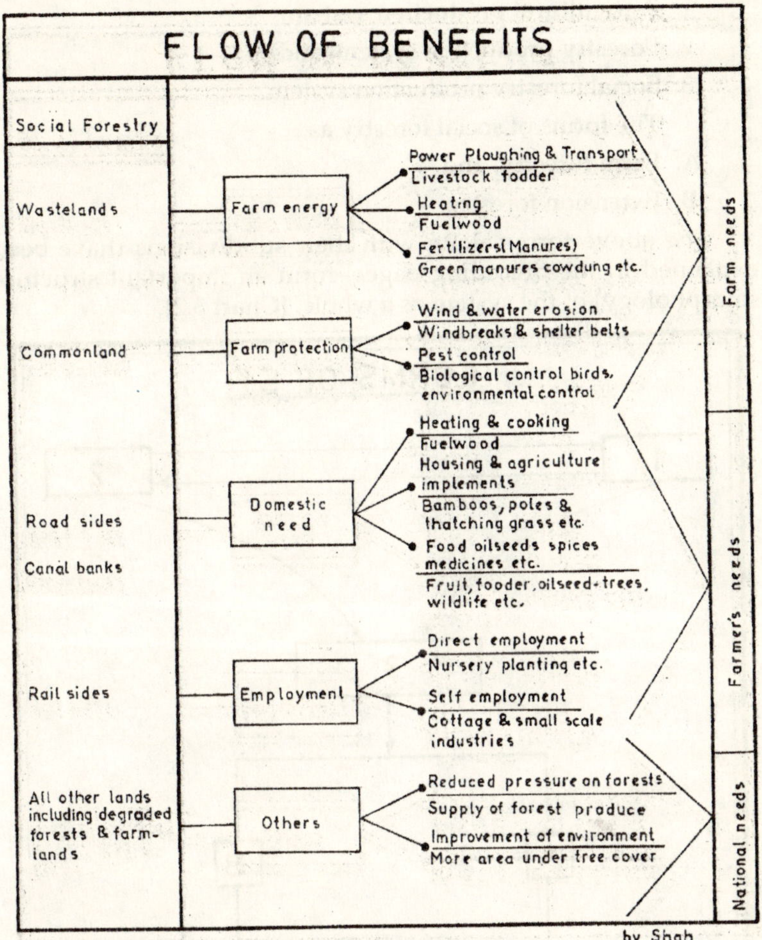

Chart : 3.1

Forms of Social Forestry

The genesis of the social forestry, actually takes root from the very concept of the forestry system. In fact forestry system is widely practiced system of agriculture, but has a separate identity of functional aspects. Therefore it may be useful to consider it a system in itself. In fact this system has a productional characteristics, therefore it is also considered as a production system. In this system there are three important components *viz.;*

1. Agricultural production system,
2. Forestry production system, and
3. Social forestry production system.

The forms of social forestry as,

A. Farm forestry, and

B. Extension forestry.

The above forms along with their subdivisions (have been explained in the preceding pages) form an important structure (morphology) of the system as a whole, (Chart 3.2).

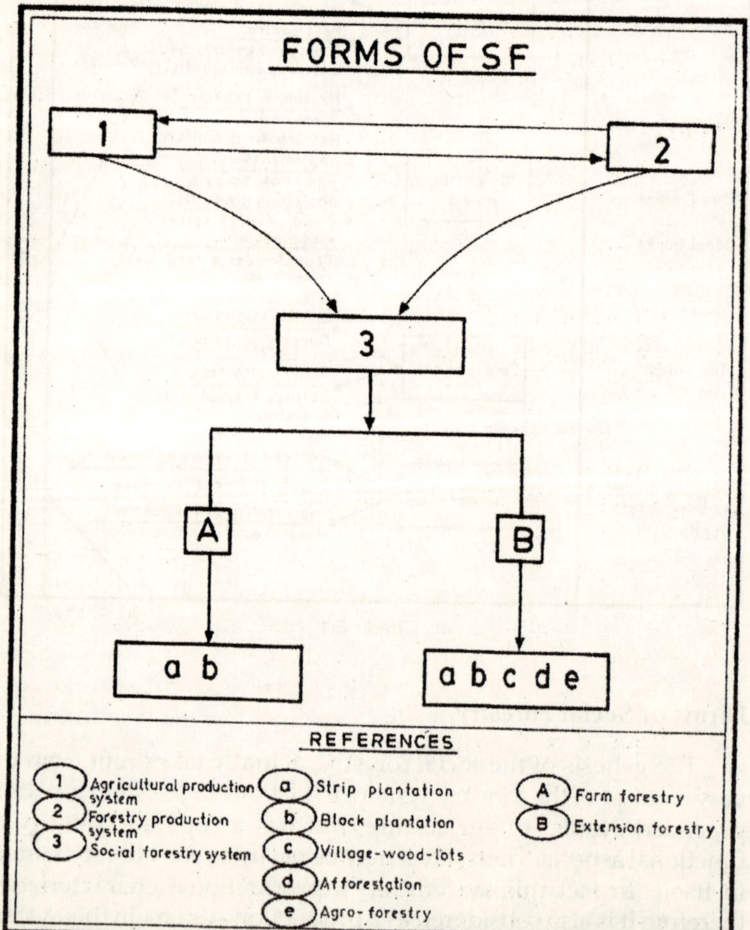

Chart : 3.2

The major characteristic features[10] of the social forestry are :

(*i*) Involvement of the beneficiaries right from the planning stage.

(*ii*) Satisfaction of the beneficiary, based on the consumption goods.

(*iii*) Community lands are used.

(*iv*) Mixed production systems *viz.*, grass, fodder, fruits, fibre, etc., are followed.

(*v*) Government control is minimal.

(*vi*) Finance is contributed through Panchayats, as it draws Government subsidies, and voluntary contributions etc.

The different components of social forestry are stated as below :

(*a*) Strip plantation along roadsides, canal banks, and rail lines.

(*b*) Block plantation in tank beds and foreshore lands.

(*c*) Village woodlots on community and government waste lands.

(*d*) Afforestation of degraded forests.

(*e*) Agro-forestry on marginal and submarginal farm lands.

Environmental Set up and Social Forestry

A world eminent climatologist Kandrew once stated that "Give me a leaf of a tree and I will tell you everything about the nature". These words have a greater relevance in the present context. Infact a tree speaks everything about the natural conditions-climate, relief, soil etc. etc. In other words the vegetation is the true reflection of the environment, *vice-versa,* environmental conditions determine and control the growth and the development of plants. Thus it may be understood that the plant and environment are two sets of a milieu, which was inter-relationships. The plant community and species, and their distribution in different parts of the earth is very largely due to the different environmental set-up in which they flourish and develop.

Therefore, it would be appropriate to say that the social forestry has a different dimension; growth and development in different parts. The present attempt is thus may be looked upon as a very small unit but has a greater relevance in the present day context.

The district as a unit of study may be a small but it has varying nature and character of land topography, climate (seasons) and soil which have provided a wider range of plant species and plant community. In this context it would be relevant to mention that the forest cover has widely been removed by the local inhabitants, resulted in degradation of environmental conditions.

References

1. F.R.I. Publication. 1968, Glossary of Technical Terms, Dehradun.
2. Shah, S.A. 1988, Forestry for people, p. 12.
3. *Ibid*, p. 12.
4. RAPA Report, 1983, Community Forestry, p. 103, Bangkok.
5. Tiwari, K.M. 1983, Social Forestry in India, Natraj Publishers, Dehradun.
6. Westoby, J.L. 1968, Changing Objects of Forest Management. Paper presented to the IX Commonwealth Forestry Conference, New Delhi.
7. Shah, S. A. 1988, *Ibid*, p. 13.
8. Anonymous. 1973, Government of India, National Commission on Agriculture, Interim Report on Social Forestry.
9. Cadelina, R. V. 1983, Community Forestry, RAPA, p. 104.
10. Vyas, N. N. (*ed.*) 1981, Social Forestry in Tribal Development, pp. 10-11.

Chapter-4

Environmental Impact Assessment

Environment and Plants

The environmental conditions very largely determines the plant growth and distribution to a level of some of the characteristics of the plants themselves. Plants vary in appearance and physiographical conditions in requirement for growth and development and in ability to tolerate environmental conditions, which are often closely inter-related.

Thus a group of plant recognisable on the earth's surface forms a part of the forest. Thus forest is in other words with its different species in which variety of plants form a group together in natural conditions, is called forest. The smallest and most widely recognisable group of plants that can be differentiated on the basis of form, or morphology is "species", the basic unit of organic classification. Basically a species is a population of individual organism, which can distinguished from the individuals of another population, in terms of three criteria.[1]

(*i*) Form or external appearance,

(*ii*) The ability of member of species to interbreed; and

(*iii*) Eco-logical requirements and tolerance of external environmental conditions.

From the above mentioned criteria for the growth of plants and development, the third criteria makes an important point while studying forests, or in other words the natural vegetative cover on earth's surface.

Morphologically distinct species can live together as their ability to make different use of the same environment. In the process of evolution some species have been disappeared, and new species have emerged, simply because of untolerable environmental conditions. Slight variation usually exists between them in characters such as height, number of branches, size and shape of leaf etc. These variation can and do arise as a result of differences in the environmental conditions under which the plants are growing.

Some species are larger, taller and more vigorous than those of the same species on poor sites. Variations in the form of coastal or mountain representatives of a particular species may be inherent genetically determine characterstics which may be retained even if the plants or seeds transplanted in different and more favorable habitats.

Many species which occur over a wide range of varying environmental conditions must inevitably compose of a varying number of ecological races or ecotypes, adopted to a slightly different habitat[2] such ecotype probably represent the most recent products of evolution.

They will tend to emerge when a species spreads into a slightly different habitat or therefore some reason such as climatic changed conditions may alter in all or part of the existing habitat. Thus the ability of the individual of a species population to exists and maintain themselves in a particular area will depend on those conditions, necessary for their growth.

Plants by and large vary not only in their essential requirements but in what is called their tolerance of environmental conditions. Any environmental condition beyond that it can tolerate either a deficiency or excess of any habitat factors; climatic, edaphic or biotic, will limit in growth and hence the area it is capable or occupying. Tolerance however varies not only between species but within a particular species in respect of different habitat factors[3].

Thus the distribution of plants is directly governed by the three basic factors;

(*i*) Climatic,

(*ii*) Edaphic, and

(*iii*) Biotic.

The above factors in other words constitute environmental conditions or in a specific term habitat for the plants. Therefore any change in any of these factors directly effect the plant growth and distribution. Hence the change in plant species is a result of change in environmental conditions. In such conditions a species may have a wide range of tolerance to temperature conditions or tolerance to soil acidity or alkalinity may be very narrow indeed, in the latter case plants tolerance may be so limited as to provide a sensitive environmental conditions.

A particular species can only occupy those parts of earth's surface (habitat) where environmental conditions are within its range of tolerance. It does not mean however that ecological conditions are unfavourable.

The present distribution of various plant species over a wide range of areas is a result of environmental change that have taken place during the course of species evolution and migration. Change in distribution and growth has been the result of changes in climate and environmental conditions during the course of their existence.

The agent of these changes is man and his expanding efforts on plant evolution and distribution, which have been so pronounced as far reaching on any physical factors operated in the past. But in a general geographical distribution of all kinds is grouped into three major categories as they are profoundly influenced and directed by the range of physical factors as a case of tolerance.

1. Xerophytes (plants resisting drought)
2. Hygrophytes (plants resisting water)
3. Tropophytes (plants of varied adaptations).

Thus it would be interesting to see that in different environmental conditions various plant species are akin to develop. But, the guiding factor for all these conditions is the environment, which is constituted by various elements of nature like water, soil, drainage, etc. alongwith the temperature and precipitation at a place. The environmental influence on plants is discussed as under.

Environmental Factors

As it has already been discussed the role of environment in controlling life cycle of plants and all other living organs including man. The climatic elements are the most important constituent factors that influence the plant life. The climate of a region is determined by its temperature, precipitation and the intensity of sunlight available and these in turn depends upon—

(1) Latitude, (2) Altitude, (3) Direction and nature of prevailing winds, (4) Direction of mountain ranges, (5) Slope of the land, (6) Evaporating rate, (7) Position of the region on the continental land mass, (8) Direction and nature of ocean currents, (9) Hours and intensity of summer.

The evaporating power of air is a very important factor in plant life for transpiration. Plant must strike a balance between the amount of water they can take up from the soil and the amount they loss through transpiration. If the water loss through transpiration exceeds the water taken in through the roots, the plant will any may even die. Xerophytes, Growing in dry arid region have a special adaptation to reduce transpiration.

The most important among the factors, water, soil and various conditions related to plant life are discussed here in the following way.

Water as a factor of growth

Water is not merely a chemical composition of H_2O in scientific terminology, but in its natural conditions, physical properties, types, quality and its movements etc. exhibit a complex relationship.

Water is found in various aggregatory conditions and in different forms. It is found in gaseous, liquid, solid and in moisture conditions which occur in different forms in different timings. Thus, it is a most important factor for development and growth of plants and fertility of soils.

The atmospheric water is received in the form of precipitation (rainfall), affect plants directly, but this water is generally found in mobile conditions in various form and responsible for the plants growth and development. In natural process water attains aggregatory to microscopic form and except influence on the generic process of plants.

All plants have certain environmental requirements that must be met, if they are to survive in due course of time. That is why, specific plant species are found with environmental adaptability in different parts of the world. Some plants are found in very moist climate known as 'Hygrophytes'. Those adopted to drought are called 'Xerophytes', and plants with the ability to adjust themselves with the existing conditions are known as 'Tropophytes'. But above all plants some way or other thrive on water. Water is the crucial factor, varying significantly in volume from place to place and in time to time. It provides the medium for the process of plant growth in the form of moisture and transpiration and thus, soil play vital role in making this process a complete cycle for the generic development. If the water demand is not met, due to evaporation from soil and transpiration from plant surface, the wilting process may go in advance and eventual death occurs. Thus, it is the temperature and precipitation which set the conditions for the growth more suitable or unsuitable. As we know that drought is a condition in which water scarcity is increased and plant die in need of adequate moisture. This sets the need for alternate source of water or abandon of agriculture. On the other hand, artificial supply of water is the only answer to this fact, if it is made available.

Of all the characteristics, form of the water movement makes a point. It is associated with gaseous, solid and liquid forms in which it enable plants to survive. Thus, the movement of water determines the quality and type of plants in different shapes and sheds. The most prominent role of water is found in liquid form which runs through surface and underground layers and is managed and controlled for crops and plants.

In spite of all that each plant family has its own water requirements as it differs in growth cycling of a plant. In a complete vegetative period of a plant various forces contain to accomplish photosynthesis and metabolism processes of plant genesis.

Edaphic Factors

Soils are composed of many mineralised elements like magnesium, aluminium, iron alongwith calcium, potassium, and phosphorus etc. Thus, it is formed of many chemical and mineralised substances and hence, reacts or behaves in accordance with its chief characteristics. The effect of these composed elements are invariably seen when they are applied through any medium of soil with water. Some of the effects are expressed as under;

1. Compound of calcium may be pure like felspar contain alumina and sllica. The effect of calcium on plant is chiefly indirect in its influence on soil structure and in making other salts available.
2. Potassium (k) affects leaf development, the process of carbon assimilation in leaves, absorption and transpiration. Potassium salts are very soluble in water and places where potash, felspar is found. Moderate rainfall is very favourable for this aspects of plant growth.
3. Phosphorus seems to influence root and seed with developments. Phosphorus compounds accumulate in the green part of plants, which is important for stock raising as calcium phosphate is necessary for bone building.
4. Iron dissolves out from many rocks and soil constituents ionised compound.
5. Nitrogen compounds-plants obtain nitrogen supply from the nitrates in the soil, through pre-existing organic matter dead plants, dead animals which are the main source of supply of nitrogen salts.

Role of Soil in Plant Life

Next to the climate soil is the most important factor governing plant life. Soil in the upper layer of the earth's crust, which exists above the hard rocks and in which plant grow, is composed of disintegrated rocks with a mixture of organic remains. As plants take up all the required salts from the soil alone, it play an important part in plant life. Factors connected with the soil are known as edaphic factors and they depend upon its chemical composition. Its physical properties, its water air content and on the presence of certain organism in it.

Primary, soil is created by the weathering of rocks. But other factors play an important role in its final constitution. Plants and animal remains form humus which enriches the soil. Some soils are sandy, porous, light to work and easily penetrated by roots, but they are dry, clayey soils retain water but they are poorly airated, sticky when wet and hard when dry. Loamy soil is composed of a mixture of clay, sand and decayed vegetable matter is the best for cultivation.

The essential constituents of soil are—

1. Mineral, 2. Water, 3. Air, 4. Decaying organic matter, and 5. Living organism.

Soil is probably the most important of man's natural resources. However, until recently it has been studied either as one of the environmental variable the adaphic factor, affecting plant growth and distribution or as an integral component of a sub-system of terrestrial environment.

Certainly the soils differ in their terrestrial ecosystem in several respects. Apart from a few specialised surface living algae, the primary biological producers (the autotrophic organisms) play an insignificant role in energy flow. In most mature above ground terresterial eco-system the greater part of the annual primary production enters the detrial food chain in the soil. Only under intensive grazing does a significant amount pass via herbivores into the terrestrial grazing food chain while the process basic to above ground system is primary biological production (or photosynthesis) that in the soil as decomposition. In the former, solar radiation is converted into potential food energy, in the later food energy is continuously used and finally dissipated as heat. However, despite the differences in energy source the soil shares the following basic characteristics in common with all other ecosystems.[4]

1. A set of interacting biotic and abiotic component.
2. Energy flow through the system.
3. Continual biogeochemical cycling of organic and inorganic elements.
4. A self generated self regulated development towards mature climax or study state conditions.

In any given soil the abiotic components of mineral matter, air and water, together provide the physical habitat for the soil organism.

However the ecological significance of all the soil bacteria is related to the fact that, firstly they constitute the important group of general purpose decomposes, secondly certain species of bacteria are practically entirely responsible for 3 vital soil processes; the oxidation of sulphur compounds, the fixation of gaseous nitrogen and the nitrification or release of inorganic nitrogen from organic compound.

Eco-System

In different environmental conditions, broadly speaking a set of natural conditions consisting of temperature, rainfall, soil vegetation etc. form an environment. Whereas the sub-conditions limiting with environmental elements form an eco-system. Thus environment is a sum total of physical and biological conditions. Whereas eco-system is a condition inherent with changing perception of temperature and rainfall or soil etc. Thus there can be many more eco-systems, in a set of environment *e.g.* tropical, temperate, or subtropical, while there are much more eco-systems depending upon the range and variability of elements. These may be of grass lands termed as grass land ecology, of forest or forest ecology and so on. In fact ecology is a term used for the micro area or sub area which have minor susceptibility of natural conditions like temperature and rainfall. On changes of temperature and rainfall, the conditions also changes for the sustainable plant and animal life.

The ecology in specific sense is invariably used as a habitat. The habitat is a set of conditions governed by the climatic, edaphic and biotic factors. In general usage habitat is described as the environment, which it has facets. But in reality it denotes the climatic conditions, soil and vegetative cover. Thus environment is a sum total of these conditions.

To further describe the environment as a factor of change of plant and animal, acts as a parameter (variable). Climate is an aggregrate condition of temperature, rainfall and humidity. Thus it constitute a set natural environment comprising of temperature, rainfall, soil conditions, vegetative cover etc. If there is any little change in any of these elements it affects the whole, and chain reaction is apparently witnessed.

Thus to understand a eco-system one has take care of these changing conditions in a region. Therefore eco-system may be described as a set of elements associated with the natural conditions; either with climate, soil or vegetative cover.

Forest Eco-System

In the development of forestry system the above mentioned environmental conditions on the whole, and its elements and sub elements in particular have a special role to play. Therefore the range of variation of plant species and community formed on different parts of the earth, is the result of these micro level natural conditions which have different ecology.

The vegetal cover of the earth surface may be described as the association and contribution of plant families. Thus vegetation form the certain combination of plant species which grow together. The life sustaining capacity of plants very much depends on the environmental factors, which combine together form different eco-systems of plant species. Thus forests and plants are no more a separate entity of a ecological domain but form a composit environment in a set of micro eco-system regions.

In fact forest cover sets conditions of plant growth and development through controlling temperature and percipitation. Attracting humidity and checking evapotranspiration, and also provide humus to the soil, which helps in plant growth. Thus the atmospheric balance of various elements is maintained through forests. But at the same time if forest cover is removed, they effect in several manner, causing overall destruction of the eco-system. Thus the vegetative cover on the surface whether in the form of natural forests or plants by man both considered a part of flora, are controlling elements of the eco-system of a region. Forests may therefore be regraded as—

1. Natural canopy of vegetation, and
2. Resource based economy (Social Forestry).

So far the natural canopy of vegetation is concerned forests are crucial and vital part of the whole system of the life sustaining elements. As it has been said earlier that forests as wild plants in the form of wood lands are the parts of the natural canopy. They are the part of biologically complex and productive form of all terrestrial eco-systems.

The ecological dominance and economic significance of trees to a great extent is a function of their size and longevity" (J. Tivy, 1982)[5]. Tree species are reckoned with their size, longevity and growth, and hence productive in regeneration. All these characteristics mark forests as an economic asset to the nation. But besides all that, their role in preserving a perfect eco-system is far greater than anything else.

By virtue of a definite structure of plant physiology, the amount and the distribution of living and dead plants (woody forests) material, which form an economic aspects too, dependent on a number of interacting elements *viz.* Species, age, types etc. Thus, a composition and structure of forests and wood lands are dependent elements in a similar complex of environmental factors as is the plant biomass and the resulting diversity of plant communities.

Plant Morphology

The principal characteristics must commonly known to botanist or biogeographer in identifying a plant species from another and distinguishing them into plant communities, is the plant morphology.

Plants vary in appearance and physiological constitution in requirements for growth and in ability to tolerate environmental conditions. Basically a species is a population of individual organism (plant or animal) which can be distinguished from the individuals of another population.

In order to acquire these conditions plants are adapted in accordance with the geographical environment. Thus, these variations can and do arise as a result of differences in the environmental conditions under which plants are grown. Those in more favourable habitats are tend to be larger, taller and more vigorus than those of the same species which occur over a wide range of varying environmental conditions must invariably be composed of a varying number or ecological or eco-types. In a wide range of a single set of a plant species is a eco-type. Those groups of plants belonging to a particular species which possess such genetically determined variations in neither form a physiology under varying habitat conditions are known as ecological races or eco-types (J. Tivy, 1981)[6].

Plant Communities

Plants are said to possess a structure; the set of life forms found in a plant community and their spatial arrangements, constitute the primary morphological features of that vegetation. A structure of a community in fact, reflects three components; *viz.*

(*i*) The vertical arrangement of the plants,

(*ii*) The horizontal pattern of spatial distribution, and

(*iii*) The abundance of each species.

Ecological Niche

In plant communities particular structure is an important element in identifying one group to another. Actually it refers to the population structure of a ecosystem, termed as ecological niche. Odum defines the ecological niche in functional terms as the status of an organism within its ecosystem. This status results from species adaptations, physiological responses, and behaviour. Whittakar extends this definition by describing each species is an eco-system by having its own place in vertical and horizontal space. (Nigel Pears 1977)[7].

Thus ecological niche refers to the prominence and domination of one particular species in overall plant communities.

Dominant species

Within the structure exhibited by a plant community one species will usually be found strongly influencing the environment of the other species present. This plant is usually referred to a dominant one. Thus, the term dominant can be expressed as a strong influence of a particular species on the structure of the community. *i.e.* Oak or Birch express influence over the other's domination in the high Himalayan region. Dominance implies competitive ability of plants.

Structure of Eco-system

From the above discussion one may come to the conclusion that there exists a complete eco-system in which every individual organism of living or dead world is interlinked to one another. It may be said that in all eco-systems the living organism interact with the non-living substances (inorganic and organic) to produce an exchange of materials. In functional terms eco-system has two functional components *viz.*;

1. The autotrophic (self feeding) components.
2. The heterotrophic (other feeding) components.

Thus, co-system possesses four basic elements :

(*a*) Abiotic substances, the basic inorganic and dead organic compounds of the habitats.

(*b*) The producers, which are mainly autotrophic green plants.

They stand as the intermediaries between the inorganic and organic world.

(*c*) The consumers or heterotrophic organism are mainly animals including man, which have intake organic materials.

(*d*) The decomposers. Or micro-organism.

The organic world of vegetation and forest canopy is an important and integral part of the whole spectrum. Because organic world by itself is governed by the external as well as internal forces (energy flow) generated by themselves. Thus, it would be appropriate to have a look on some of the elements responsible for the forest eco-system.

Forest Resource Based Economy in an Eco-system

Forests are considered as the great asset of resource and hence a great economic value in the form of woodland areas. Their economic significance lie in their use.

Man's perception about the forests has been a different one. Though relationship between man and forest has been ambivalent. But, at one time, forests have been looked upon as a potent enemy, a barrier, impending ease and rapidity of movements, obscuring wild animals etc.

But on the other hand, how forests can be a great asset, a resource to human being, has lately been realised. They are in true sense, a great protector of the human life on the earth and living organs including itself (vegetation) preserving the eco-system and maintaining balance between nature an man's harsh attitude from time to time.

Though, very lately, the importance of the forests as a resource in economic terms, was realised with a great danger of the human civilization at brink. Because, with the span of development, man exploited this resource recklessly by clearing dense forests trees from all the parts for his barring needs, inviting the natural catastrophe. The rise and fall of human civilization have been accompanied by systematic and deliberate clearing of forests. As a result existing forests and wood lands have become increasingly concentrated on one part or the other barring growth of trees.

The nature and value of forest resources depend on the type

of forests. Forests in general provide the primary need of biological products as fuel, fodder and fruits. For the better understanding, we may provide the following categories, through which we fulfill our needs, they are;

1. Agro Forestry
2. Social Forestry, and
3. Silvi-culture

1. Agro-Forestry

In the competing world of needs, men has acquired multi-dimensions of exploitation of natural resources. Trees make a relatively meagre contribution to total food supplies at present. The use of forests as a source of fodder for domestic animals, under present systems of management, conflicts with a handicapped by that for wood production. Forests have long been a source of forage for animals. Thus, agro forestry serves dual purpose of economy.

In this system the forest trees and agricultural crops which can be grown together can be gleaned from several sources. Blanford (1958)[8], has pointed this practice as *'Taungya;* located in Burma as back as 1856. In the tropical region the forest plantation was founded on this principle. King (1968)[9] and Von Hesmer (1966, 1970)[10] have reported about the tree and crop cultivation together, and found no adverse effect of crops on trees and *vice-verse.* There are reports that the way of doing agro-forestry yields more. Agarwal, *et. al.* (1976) found a better harnessing conditions specially under improved plantation under Khejri Singh and Lal (1969)[11] are of the opinion that it improves the soil fertility, through ascribing the input of microbial fertilizers in the soil. Khejri (prosopis cineraria) is a true species abundantly found in arid and semi arid parts of Rajasthan alongwith fields.

A good example of agro-forestry may be cited from the moong and guar crops in the inter row space of the prosopis cinerasia tree (lopped and unlopped) in Wn. Rajasthan. Moong and guar both are a varfiety of pulses, but provide good fodder for the animal also. They are very much preferred in the dry areas of the Wn. Part of Rajasthan. The economics of these alternative land use systems has been worked out by Gupta (1980), that the net returns per ha. Under annual crops is quite substantive.

In the agro-forestry system the more interesting feature of farming activity has been animal ranching alongwith some crops which provide the forage to the animals. Thus, intercropping system has developed alongwith tree plantation. These practices are well known as 'Silviculture and cropping silvipastoralism etc.

Although, modern silvicultural techniques have developed slurly but their origin has been an ancient one. It is a process of regeneration of plant micro-organism. The coppicing method is dependent on the ability of many hard wood species to produce a number of newstems from the stumps or stools of felled trees. The stems can then be harvested on a rotation depending on the species and size of wood required.

On the contrary, the silvipastoral system is related to livestock husbandry in which fodder tree and shrub species are grown for the forage and shelter to the grazing animals simultaneously. Zizyphus nummularia and prosopis cineraria are important forage tree species which grow in cultivated fields without any detriment to the crops grown in association with it. The main object of the silvi-culture is to manipulation of forests and wood lands to modifying and control local and micro-climatic conditions in the interest of other types of land use. In this case, the main purpose is to provide shelter for agricultural crops, domestic livestock, buildings and to develop highway avenues.

On a large scale, the effect of the shelter belts is to reduce wind speed, with a consequent reduction of evapotranspiration and an increase in humidity and temperature. Though, in variable decrease in crops yields in the immediate vicinity of the forestedge as a result of shading and root competition is also reported. But, it is compensated with the higher yields, than the unsheltered areas. Thus agro-forestry helps in raising the productivity of crops.

2. Social Forestry

Although, there is no marked distinction between agro-forestry and social forestry, as has been discussed above. But, so far the social needs are concerned, mainly confined to the food, fodder and fuel, the forest becomes a part of SF as an important source of economy of the people. It has been discussed elaborately in chapter 3.

3. Silviculture

It is the art and science of cultivating forest crops[12]. The study of the life history and general characteristics of forest trees and crop, with particular reference to environmental factors, as the basis for the practice of silviculture. In other words silviculture commonly refers only to certain aspects of the theory and practices of raising of forest crops. 'Forestry' the true equivalent of agriculture and silviculture is used to imply only the methods of raising tree crops and their growth and care up to the time of the final harvesting. A general term for all operations carried out immediately after the main fellings, besides cultural operations are called silvicultural operations. Where as silvicultural system is a method of silvicultural procedure worked out in accordance with accepted sets of silvicultural principles by which crops consisting forests are tended, harvested and replaced by new crops of distinctive forms[13].

Changes in Environmental Conditions

Any change in the nature's course is regarded as a change in environment. And a change in environment is considered a human impact on nature's way that means human interference in the nature. Therefore Andrew Goudie has reghtly quoted Yi-su-Tuan (1971)[14] that' the fact of diminishing nature and human ubiquity is now obvious'. That means the human interference in nature's process is always considered as an encroachment. This interference may be for the development of human society or for the well being but in some way or the other, in a long run, it leads to a degradation of environment.

The environment is composed of many components of natural conditions, as it has been discussed in the foregoing pages, like climate, soil. Vegetation, relief etc. Therefore if there is any change in any of the above elements it brings, change in the overall structure.

On the contrary, the human efforts in order to perform better living, he always acts in a particular manner. This act is termed as development. Development in real sense is nothing but a process of human efforts for the better living. Thus, this process directly or indirectly involved men to interfere in the natural process. By doing so, he achieves a living standard of his choice which termed as the development.

This sort of development in the form of human interference marked as the human impact can be seen in various fields on different levels. Thus human impact here in the present context is seen as the plant and man relationship.

It is not to emphasize this fact here that how plants (forests) affect the human life or what is the impact of plant on human being. But it makes an important point in the light that how man affects the plant life in general. This point brings out clearly the change caused by the man in the nature's way. Because by cutting down trees or forests on a large scale cause changes in the temperature and rainfall, which in turn bring changes in the climate, hence the change in the total environment or the eco-system of a region.

The change in the climate affects the overall change in human conditions and their associated activities may be of cultivation of crops, forestry or live stock breeding etc... etc. The soil erosin, water logging, down cutting of trees, desertification etc. are the areas directly affected by the any change in the climatic conditions.

Thus the changing environmental conditions may be better understood in the form of human impact on vegetation, human impact on soil, human impact on water, human impact on climate etc. as very well described by Andrew Goudie in his book - the human impact on the natural environment (1981 and 1990 third, *ed*.).

Human Impact on Vegetation

The human impact on vegetation with particular reference to the forests and the forestry systems is the result of the human interference. In other words the forests have been removed with the span of human cultivation on a large scale, compelled man to protect forest wealth in other words the environment. Otherwise what has been the reason for developing different systems of forestry like :

1. De-forestation
2. Re-forestation
3. Afforestation
4. Agro-forestry
5. Farm forestry
6. Social forestry

If we go into the details of the above forms of the forestry system, we come to the interesting point that the following functional forms may be categorised.

1. Degradation system;
2. Conservation system;
3. Production system.

In the first form or system we take the degradation of forests. As it is well known that at the beginning of human civilization forests on the large scale were exploited for the human needs, more particularly the fuel. The history speaks of the wide spread removal of forests from the selected sites *e.g.* Run of Kutch.

It is clearly understood that with the removal of the forest wealth the environmental changes were witnessed in different parts. Thus the degradation process of deforestation started with the enroot of the tree cover.

The second system of conservation of forests was adopted mainly because of the deforestation. Deforestation needs reforestation or afforestation. Both these systems have a particular nature of conservation. This system infact is a regeneration of plants which is adopted at two levels *viz.* One, natural regeneration and two, artificial regeneration. Reforestation and afforestation systems belong to the latter category of process.

Reforestation is a process accomplished where forest cover is removed/destroyed, it is restocking of trees on sites which originally had trees. Reforestation is done for gap planting by (*a*) supplementing natural regeneration in failed or backward patches, (*b*) replacing natural regeneration to augment scanty, slow or uncertain natural regeneration, (*c*) restocking blanks caused by fire, frost, mining, (*d*) introducing successful exotics such as Israeli Babool or Safeda.[15] Whereas afforestation is adopted in fresh areas the areas where no forests existed.

Afforestation may best be understood as V.S. Saxene puts it in the following words—"Means the 'de novo' establishment of desired species in an area which never had trees at least for a few decades—for lack of seed trees, aridity, unstable soil, or swampiness"[16].

The third system of forestry which includes Agroforestry, Farm forestry and social forestry related to the production system as they provide or fulfils the social needs. The basic idea behind the Agroforestry, Farm forestry and the Social forestry is to obtain fuel, fodder and fruits.

From the above explantations we may come to the fact that in different areas in different environmental set up and eco-system we find different forestry systems, enable us to protect environmental changes. The above mentioned forestry systems provide us a wide range of vegetative cover enhancing soil productivity, soil conservation and fertility, maintaining water table, etc. on one side and checking soil erosion, de-cropping of land, flood devastation etc. on the other.

Therefore we may conclude that the environmental conditions in general and eco-system in particular for the forestry systems play an important role for their change as well as transformation. Thus human impact on nature and assessment becomes a part of intensive study to understand man and nature relationships.

References

1. T. V. Joy. 1982, Biogeography. P. 110, Longman, London.
2. *Ibid.* P. 133.
3. Nigel Peass. 1977, Basic Biogeography. P. 83 Longman, London.
4. T. V. Joy. 1982, Biogeography. P. 380. Longman, London.
5. *Ibid.* P. 258.
6. *Ibid* P. 133.
7. Nigel Peass. 1977, Basic Biogeography. P. 83, Longman, London.
8. Blan Ford. 1987, Dryland Agriculture in India. P. 13, Longman, London.
9. King, K.F. S. 1968, Agri-Silviculture and Taungya System, Bulletin Vol (1) P. 109, Deptt. of University of Ibadan.
10. Von Hesmer. 1970, Der Kombiniert land forest wirst-win chaftliche Anban-II, Tropicasches. Subtropiches Asien, P. 219, Ernst-Klett, Varlag Stuttgast.
11. Singh, K.S. and Lal, 1960, Ann. Arid Zone, 8. Pp. 33-36.
12. F.R.I. 1969, Glossary of Technical Terms. P. 135, Dehradun.
13. *Ibid.* P. 135.
14. Goudie, A. 1990, The Human Impact on the Natural Environment. P. 1, Oxford, London.
15. Saxena, V.S. 1993, Tree nurseries and planting practices. P. 9, Delta Printing Bombay.
16. *Ibid.* P. 9.

Chapter-5

Impact Assessment of Social Forestry

General

The impact of social forestry has several dimensions and descriptions of which may be given under the following heads; Bio diversity, human life, wild life, Agriculture, soil and water, Rural and Urban complexes, etc... etc.

In broader terms the above mentioned heads have a greater bearing in determining the environmental conditions, together with natural and social set up. These conditions, by and large affect each other and are responsible for the change. However, the changing conditions are the result of the human interference in the nature's course. Therefore, the natural environment is changing in the gradual process of human deeds. This process, though taking a long way of its own and the results are seen after several years.

Thus impact of human activity on nature and nature's on human life is an inter-dependent activity. Human response to nature's way and *vice-versa* is a factor of co-existence. The results may readily be seen in deforestation, soil depletion, flood and famine at times. Therefore, the study of impact assessment in general and the social forestry in particular is a way of inter-dependence on one side and interaction on the other.

Biotic and Abiotic Resources

BIOTIC RESOURCES

So far the biotic resources are concern in the district context, they may be classified and categorised in the following forms :

1. Flora, and
2. Fauna.

1. Flora

The flora or vegetative cover in the district is a distinct in many respects. It has a wide range of floral diversity with a variety of innumberable species. The most prominent tree species among them are — Dhokra (Anogeissus pendula), Khair (Aca. Catechu), Salar (Boswellia-serrata), Gurjan (Lanea grandis), Kumta (Acacia senagal), Babul (Acacia nilotica) Vilayati Babul (Prosopis juliflora), Ronj (Acacia leucophloea), Tendu (Diospyros melanoxylon) Anwala (Emblica officinalis), Rein (Mimosop hexandra), Dhak (Butea monosperma), Mahua (Madhuca latifolia), Neem (Azadirachta indica), Baheda (Terminalia belarica), Arjun (Terminalia arjunȧ), Khirani (Wrightia tinctoria), Karaya (Steruculia urens), Jamun (Syzygium cumini), Khajur (Phonix-sylvestris), Khejari (Prosopis spicegera), Jal (Salvadora-oleodies), Bans (Dendrocalamus strictus) etc... etc. Among grasses are Lapla (Aristida depressa), Anjan or Dhama (Cenchrus ciliaris), Dub (Cynodon dactylon), Munja (Sachra munja) etc... etc.

Besides these tree species, there are several vegetative forms like shrubs and bushes, herbs and grasses, Bamboos etc. forms an important part of the bio-mass in the district. The different categories of floral species may be seen in the Appendix-I. Description of some of the dominant trees may be provided as under :

1.1 Babul (Acacia nilotica)

General Description. It is a moderate sized multipurpose tree of leguminosae family having dark brown and much fissured bark, umbrella shaped crown. The spines are straight and white, 1-5 cm. long and leaves are bipinnate, 2-10 cm. long having 10-25 pairs of leaflets. The flowers are yellow in colour, in axillary globose heads, pods are usually solitary 10-15 cm. long, 8-12 seeded, constricted between the seeds.

General Distribution. It is indigenous to Sindh and north-western India, cultivated through-out in the drier parts of India. In Bhilwara it is very common except uppermal area. It comes on variety of geological formations but prefers alluvial and black cotton soils.

Silvicultural Characters. Babul is a strong light demander, develops a long tap-root in the drier areas and strong lateral root system in moist areas. It pollards well but poor coppicer and root suckers are not usual. It is frought hardly but frost and fire tender. There are 3 main races of babul *viz.;*

(*i*) Ac. nilotica-var. typica

(*ii*) Ac. nilotica-var vediana

(*iii*) Ac. nilotica-var cupressiformis

But for forest purposes, race (*i*) is most important. The viability of seed is good and both natural and artificial regeneration is successful.

Economic Importance. It is an excellent multipurpose tree, heart-wood reddish or dark-brown, average weight about 1000 kg. per cum.

The heart-wood is used for wheels, well curbs, oil crushers, boat, agricultural implements and in house building. It gives excellent fuel, pods serve as fodder for cattle, goats and sheep. The gum is edible and used for various purposes, bark and pods are rich in tannin.

1.2. Dhokra (Anogeissus pendula)

General Description. A small deciduous tree of combretaceae family with a short crooked bole, sometimes straight readily distinguished by its graceful pendular branches. The tree grows to a height of 13 meters and girth 1.5 meter under favourable conditions. But, in Bhilwara it is generally grown to height of about 6 meters and girth of 90 cms. Bark is smooth and silvery grey. Leaves opposite or sub opposite, 1-2.5 cm. long elliptic, usually 6-7 pairs, petiole is less than 5 cm in length. Leaffall occurs usually in January and remains leafless till may June, leaves turn reddish black before falling. It bears small flowers during September-

October, fruits ripen from December and continue till march. The viability of the seed is very low only 3 to 5 per cent though produce abundent fruits annually.

General distribution. It takes root on variety of geological formations chiefly geniss, schist, slate, quartzite, lime-stone and traprock. It require good drainage and avoid most areas grows best on loose allavial soils. It is found all over the district except northern parts of the district. It is the principal species of the region its main associates here are khair, Grewia species, Salar and in western part of the district is Kumta.

Silvicultural characters. A strong light demander, frost hardy and suffer from severe drought. It can sustain temperature ranging from 0°C to 49°C. It coppice and pollards well but the growth is rather slow. It produces root suckers freely.

Economic importance. The wood is hard, compact, tough and elastic, weight about 950 Kg. per cum. It is used for cart axles, axe handles, poles for carrying loads, also for house posts and agricultural implements. It also gives very good fuel and charcoal. It produces edible gum which is soluble in water and leaves are valued for tanning. It is an important tree for clothing dry degraded areas.

1.3. Dhaura (Anogeissus latifolia)

General Description. It is fairly large sized deciduous tree of combretaceae family. The bark is smooth whitish with slate colour patches, with white exfoliating in irregular rounded scales leaving shallow depressions. Its leaves the alternate or subopposite. 5-15 by 2-8 cm, ovate or ovate-lanceolate, apex obtuse or emarginate, glabrous when manure. Leaf fall occurs chiefly in February and remains leafless until April-May. The leaves turn coppery red to reddish brown before falling. Flowering in December-January forming auxilary recemes and yellowish in colour. The fruits ripen in February and fall soon after ripening. The tree seeds abundently every year but the fertility of the seed is very poor.

General Distribution. The tree is characteristic of dry deciduous forests but also found common in moist deciduous forests. In drier hill slopes it grows more or less gregariously, avoids water logged areas. It grows on variety of geological formations including sand stone, lime stone, metamorphic rocks and laterites. It requires good drainage and avoids swapy as well as too drier tracts. In Bhilwara district it is found in south-eastern portion **only.**

Silvicultural Characters. A light demander and can withstand slight amount of shade in early stage. It is fairly drought and frost resistant, usually it is not browsed. It produces rootsuckers, coppices an pollards well if cut at the right reason *i.e.* in March. It can withstand temperature conditions from 0°C to 47°C.

Economic Importance. The wood of An. latifolia is extremely hard, strong, elastic, and tought, extensively used for tool handles, cart axles, agricultural implements, construction of house posts and for all purposes where great shock resistance, toughness and elasticity is required. It also gives good fuel and excellent charcoal. It yields a gum of commerce, a good substitute for gum arabica, young leaves and bark is rich in tannin.

1.4. Khair (Acacia Catechu)

General Description. It is a moderate sized deciduous tree of leguminoceae family with rough dark colour bark. Its young shoots are brown or purple, spines hooked, and shining black or brown. The leaves are bipinate 10-15 cm. long, rachis prickly with a gland below the lowest pair and one between the top pair of pinnate, leaflets 12-20 pairs. The flowers are pale or cream coloured in lax auxiliary spikes 5-10 cm. long, sessile. The pods are 5-8 cm. long flat, brown when ripe 3-10 seeded and seeds are flat dark brown.

General Distribution. It is distributed throughout the India except in the moist humid regions. It is found gregariously on sandy alluvial deposits of rivers and streams and in dry types of forests on high land away from water courses mixed with other xerophytes. It occurs in a variety of geological formations, thrives best on porous alluvium comprising sand and on well drained sand stone. It is one of the main associate of Dhokra and common all over the district except northern parts.

Silvicultural characters. It is strong light demander, frost hardy but young seedlings are frost tender. It can sustain temperature ranging from 0° to 50° C. It coppices well and produces rootsuckers when roots are exposed. Though it is xerophilous in character but suffers from abnormal drought, it is being damaged by browsing and responds well to closure. Porcupines are particularly destructive, gnawing the bark off from the base and killing the trees.

Economic Importance. The heartwood of khair is dark red and very durable average weight is 980 Kg. per cum, Katha and cutch is obtained by boiling the chips of heartwood. As the heartwood is extremely hard, used for house posts agricultural implements, cart wheels, tool handles, sugarcane and oilseed crushers etc. It also gives excellent firewood, charcoal, gum of medical importance and leaves for fodder.

1.5. Khejari (Prosopis spicigera)

General Description. It is a medium size tree of Leguminosae family with pendular branches, bark whitish grey and much fissured. The branches are armed with prickles, leaves bipinnate, opposite, 2-8 cm. long having 7-12 pairs of leaflets. The foliage becomes thin towards end of cold season due to leaffall and leaf renewal occurs in February-March. The tree bears flowers in March-April and fruits ripen in June-July.

General Distribution. It is distributed all over the district except south-eastern part (uppermal) which receives more precipitation than the other parts, but it is more common in western and northern parts which are comparatively dry and arid. It is a principal species on the higher and older alluium and sandy areas in the district. Its natural regeneration is more or less confined to depressions and absent on upper reaches.

Silvicultural Characters. It is a strong light demander, seedlings are sensitive to frost and drought but once established are very resistant to both frost and drought, it has very long tap root system sometimes up to 20 feet in sandy areas, coppices well up to a moderate age and reproduces freely by root suckers. It can also sustain to pollarding. It is one of the most suited species for Agro forestry and silvi-pastoral plantations in dry and arid areas, where temperature is as high as 50°C and rainfall is 75 mm. only.

Economic Importance. The wood is an excellent fuel, weight about 800 Kg. per cum. It produces very good fodder called "Loom for camel, goat and cattle, the pulp of the pod is eaten by human being.

1.6. Mahua (Madhuca indica)

General Description. Mahua or Indian butter tree is a member of Sapotaceae family, large deciduous tree with short bole and spreading branches, grows to a height of 20 to 50 meters and girth 2.5 meter and above. It is a most important tree of rural economy as it provides delicious and nutritive flowers popularly known as Mahua flowers. Its bark grey or blakish with vertical cracks, leaves clustered at the end of the branches, 8-25 cm. long elliptic, accuminate or obtuse, leaf-fall gradual from February to April and renewal takes place in April-May. Flowering March-April fruiting ripen from June-August and falls at once after ripening.

General Distribution. It is very widely distributed in the country in mixed deciduous forests often growing on a dry rocky or sandy soil up to 1000 meter elevation. The tree is however comparatively more concentrated in the out skirts of villages than in the distant forests. In Bhilwara district it is found in Mandalgarh Tehsil and along southern boundary of the district, but very rare in western and northern parts.

Silvicultural Characters. It is a strong light demander and get suppressed under shade, generally drought and frost hardy but suffers from abnormal drought and frost. As the tree has a large spreading root system hence wind-firm, coppices fairly well but suffers from browsing by cattle. The tree suffers from loranthus attack and some times trees are killed. It can sustain to 1°C to 47°C temperature and rainfall varying from 750 mm to 1850 mm.

Economic Importance. The wood is hard and durable, weight about 1000 kg per cum but rarely felled, as it is more valued for its flowers and seeds which are often used as articles of food. A coarse spirit is distilled from flowers, which is a powerful stimulant tonic, fruits too are edible, dried and ground into meal, a thick oil is extracted, solidifies in to butter on colling and hence the name butter tree. It is used for cooking and lightningby the rural people. Medicinally the tree is very valuable as the bark is used for curing leprosy and for healing wounds. The collection of edible flowers and fruits is an important business of rural masses. A single tree can yield the revenue to the tune of Rs. 500 to 1000.

1.7. Neem (Azadhiracta indica)

General Description. A moderate sized to large, evergreen tree of Meliaceae family, with dense rounded crown and bark thick cracked longitudinally and obliquely, leaves bipinnate, 20-30 cm. long, crowded near the end of the branches leaflets 9-15, sub opposite, accuminate, bright green and shinning above. It remains leafless for a brief period on very dry localities. New leaves appear in March-April. Panicles of small white flowers, scented and honey bearing appear from March to April. Fruits ripen in June-July, is a drupe 1 to 1.5 cm. long one seeded and viability of the seed is very poor, only of fortnight.

General Distribution . A common tree on the village sites and along the foot hills. It growns on most of the soils, doing well on clay and black cotton soils. It thrives better than many other species on dry stony shallow soil, hard calcareous soil with a pan at shallow depth but avoids lands liable for inundation. It is found from lower Himalayas to Kerala and Kutch to Assam. In Bhilwara it is found wild all over the district.

Silvicultural Characters. Neem is a light demander, but can push itself through thorney bushes. It is drought hardly but extremely frost tender, coppices well and produces root suckers. It is grazed and browsed but can also come up in spite of frequent grazing damage. It can sustain temperature ranging from 0°C to 45°C and rainfall 300 mm. to 1150 mm. As the seeds fall during the rainy season, germinates well and spreads over wide distances. The birds eat the fleshy fruit and serve as means of dispesal. Its capacity to come up and establish under difficult conditions is a further factor facilitating its expansion.

Economic Importance. The wood is reddish brown when dry, used for house building and agricultural implements. The wood weight about 1100 kg. per cum. The bark is bitter and used in tanning and medicine. Every part of tree *i.e.* bark, gum, leaves are also used as fodder and the smoke of the leaves is mosquito repellent. Its seeds contains about 40 per cent oil, which is used in lightening, medicine, chemicals, insecticides, and soap industry. Neem is called "Ghar Ka Vaidhya".

1.8. Tendu (Diospyros melanoxylon)

General Description. It is a small to medium sized deciduous tree of Ebenaceae family with grey branches, bark exfoliating in regular rectangular scales. Leaves often sub-opposite, coriaceous, 8-20 by 2-5 cm. Elliptic or elliptic oblong. The tree is leafless for a very short time in the March but seldom quite leafless, the new leaves appear in April with flowers and the fruits ripen from March to May of the following year. Fruits yellow when ripen.

General Distribution. It is a most characteristic species in dry mixed deciduous forests throughout the India and occurs abundantly in Madhya Pradesh, Maharashtra, Orissa, Andhra Pradesh and sparingly in Rajasthan, Gujarat, Bihar, West Bengal, Tamil Nadu, Karnataka etc. In Bhilwara mostly it is found in Mandalgarh tehsil the other parts generally do not have this species.

Silvicultural Characters. It is a strong light demander but withstand moderate shade in the early stages. It is frost and drought hardy, withstand severe frost and drought conditions than any other species, seedlings are susceptible to excessive moist or waterlogging conditions. Normally it produces a long stout taproot system up to a depth of 2 meter and lateral roots run horizontally up to a length of 10-12 meters. It is wind firm and coppices well, but grow slowly, produces vigorous root suckers, and immune to damages from grazing and browsing by cattle and wild animals. It can sustain temperature conditions ranging from 1°C to 48°C and can come up in areas which receives rainfall from 500 mm to 1500 mm. Natural regeneration by seed and root suckers is plentiful. Silviculturally the tree is very important in afforestating dry, poor soils and rocky hills due to its hardness capacity to withstand any maltreatment.

Economic Importance. The tree is commonly known as 'tendu' and has remarkable economic importance. The leaves are an important minor forests produce of great commercial value owing to its extensive use in the manufacturing of "biddis". India is the biggest bidi producer and consumer in the world and a good source of earning foreign exchange by exporting biddis. Besides this is known for its black heart wood known as 'Indian Ebony'. Which is used for carvered waliking sticks, picture frames and fancy articles. Wood is also used for the construction works, agricultural implements and as a mine props etc. The fruits are edible and readily eaten by poor people.

1.9. Vilayati Babool (Prosopis juliflora)

General Description. It is small or moderate sized deciduous tree armed with stout scattered thorn, a native of west tropical south America, where the climate ranges from the arid to the semiaid. It was introduced into India in 1877 from seed obtained through Kew proved of great importance for afforestation work in refractory areas. It grows in a wide range of soils, also in very poor soils as in dry, gravelly, shifting sands, alkaline and saline areas.

General Distribution. As discussed above it is an exotic species and was introduced in the district about 60-70 years ago. Trees upto 50 years of age can be seen along N-N Road *i.e.* state highway number 4. It can grow in a wide range of soils therefore it is found all over the district, but more frequent in the arid and semi arid parts of the district and is spreading to new areas by means of seed disperal.

Silvicultural Characters. Prosopis juliflora is extremely drought hardy but saplings are frost tender, resistant against browsing damage. It is a pronounced Xerophoyte which does not like moist conditions. It starts fruiting early *i.e.* 4-5 years, pods ripen in May-June. Seeds retain viability 1.5 to 2 years. Due to seed coat conditions, exhibit some dormancy which may be over come by treating with concentrated sulphuric acid for 15 to 30 minutes or by placing in boiling water and cooling for 24 hours. Direct sowing as well as transplanting is quite successful. It is a very good cippicer and comes well to pollarding and also produces root suckers.

Economic Importance. It is a good fuel and produce fencing material for the agricultural fields, also grown along the farm boundaries. It is also a scarcity fodder and food, the pod being sweet and edible. It is a good sand binder, hence planted to arrest the shifting sanddunes. The wood if water seasoned, durable and used in agricultural implements and tool handles. The char-coal of very good quality is prepared from the wood and the bark is used in tannin.

2. Fauna

In the category of fauna there are distinct and innumberable species of various breeding groups of animals including vertibrate and non-vertibrate. The most significant and important category of fauna may be listed as :

(i) Wild-Animals
(ii) Birds and
(iii) Reptiles etc.

The different category of these faunal communities may be seen in the Appendix II and III.

The above mentioned flora and fauna as a biotic resources play an important role in the social forestry as a whole. Some time they help in adding social forestry and maintaining eco-system of the region. From this point, their study form an important part.

Abiotic Resources

Under the abiotic-resources the most important among them are soil, water and minerals etc. in the district. The mineral resources as a direct earning source of people of the region, attracts attention of the planners and scholars. In fact some of the mineral resources have so instrinsic impact on the economy of the people and their exploitation to the extent of degrading environmental conditions of the region has direct impact on the eco-system. Because some of the mining quarries are directly located at the heart of the forest areas of the district. (Fig. 5.1). This shows that the mining activity operated for mica, which has large circulation over country, china clay, soap stone and sand stone etc., extensively have affected the eco-system of the region. Hence the tree cover in most of the parts of the quarring areas in the vicinity has been removed and exposed to the nature.

Forest Types and Eco-system

Forests are vital to the environment and hence to the eco-system of the region. As it has already been discussed in the preceding chapters that different forest covers affect people and their activities on the whole. It is also a bearing factor that forest cover, not only affect the human being but also to the natural conditions. Hence the role of forests may be visualized from the fact that different governmental bodies of the world-over has maintained the forests from time to time. Ecologists have emphasized that every country should have 33 per cent forest cover to maintain the perfect natural environment and hence the eco-system.

From this point of view government of India also has tried to maintain the required area under forests. But unfortunately in different parts of the Indian subcontinent in general and Rajasthan in particular has different story of the gradual degradation of forest cover in recent parts. With this context the Bhilwara district which forms about 7.6 per cent area under forests consists 1.27 per cent under the actual cover. The situation actually speaks about the removal of the forest cover, to fulfil the needs of the local residents. It is not a wonder to say that even the state has 9.32 per cent area while the actual forest cover consists of 3.83 per cent".[1]

So far the forests classification in the district is concerned, various authors have provided the types of the forests taking different criteria. The popular classification of forests is found as under :

1. Administrative basis
2. Vegetal basis, and
3. Climatic basis.

1. Administrative Basis

On the administrative basis generally the forests have been classified into three main categories. They are,

A. Reserved forests,

B. Protected forests and

C. Unclassed forests.

In the district Reserved Forests (R. F.) cover about 54.7 per cent of the total forests area, 34.6 per cent protected forest and 10.7 per cent unclassified forest area. This shows that reserved forest covers more than 50 per cent of the total forests where as hardly 10 per cent forests are open for the general public.

2. Vegetal Basis

On the basis of the vegetative cover largely composed of the dominant tree species, C.M. Mathur[2] has classified the forests of the state of Rajasthan, in which, the following forest types are located in the Bhilwara district (Fig. 6.2).

A. Anogeissus pendula

B. Degraded Anogeissus pendula

C. Anogeissus latifolia

D. Degraded Anogeissus latifolia

E. Mixed forests

F. Grass lands.

But for the present utility purposes the forests of the district have been classified into four main categories on the basis of the dominant tree species they are :

(*a*) Dhokra forest,

(*b*) Mixed forest

(*c*) Grass lands, and

(*d*) Degradation Dhokra forest.

Each of these forest cover constitute about 29.3 per cent, 40 per cent and 9.1 per cent and 21.6 per cent area of the district forests respectively[3].

3. Climatic Basis

According to the champion and Seth[4] the forest types are classified into VI major types on the basis of the climatic conditions *viz.*

I. Moist Tropical Forests,

II. Dry Tropical Forests,

III. Montane Sub-tropical Forests,

IV. Montane Temperate Forests,

V. Sub-alpine Forests, and

VI. Alpine Scrub.

But in Bhilwara district only 'dry tropical forests" are located which can be further classified into;

(*a*) Tropical dry deciduous forests, and

(*b*) Tropical thorn forest.

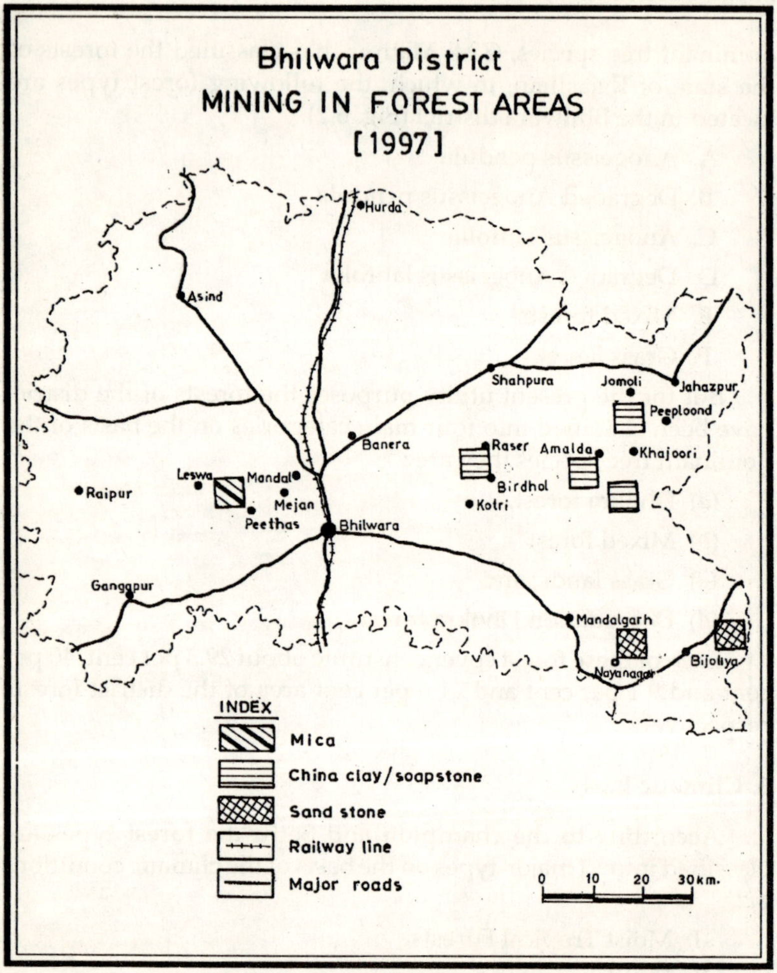

Fig. 5.1

The above classifications as provided by the C.M. Mathur and Champion and Seth may not be treated as the exclusive final classification as they are intermingled with climate and tree species, but for the general distribution purposes they may be considered important for the present context. It is also noted that the area formerly dominated by one particular group of trees is later disappeared and lying open. Thus the types located earlier exist nowhere, now.

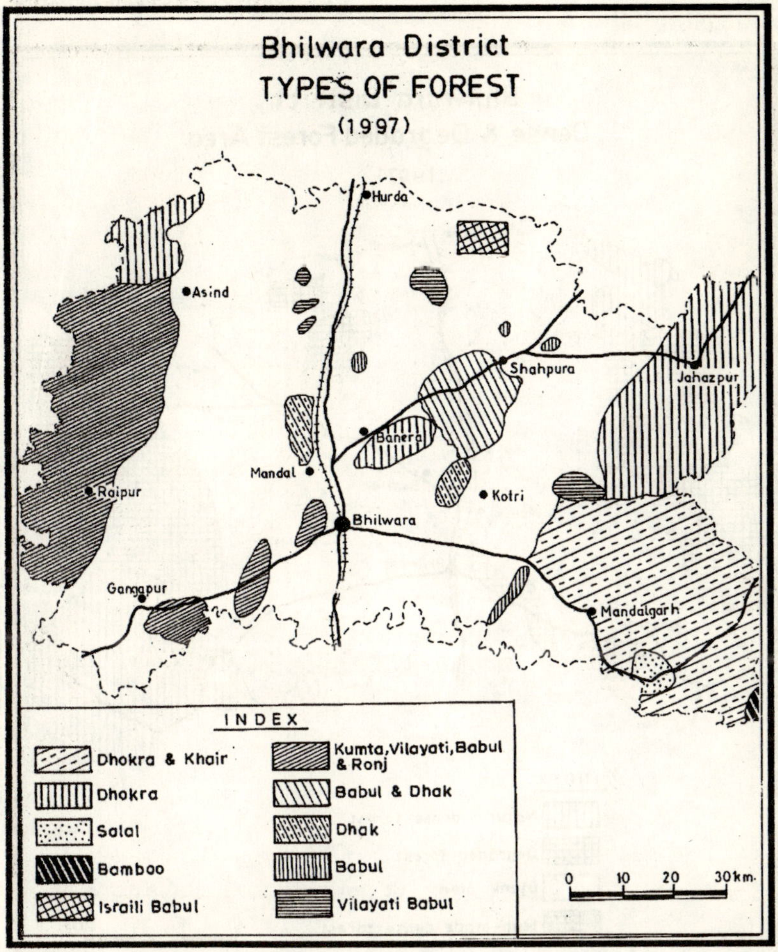

Fig. 5.2

Therefore, for the present context the forest cover in the Bhilwara district is better considered on the basis of the density of vegetation. The criteria for the density taken as follows (Fig. 5.3).

1. Dense forests > 0.4
2. Degraded forests > 0.1 but < 0.4, and
3. Blank area < 0.1

The above discussed various factors of natural conditions,

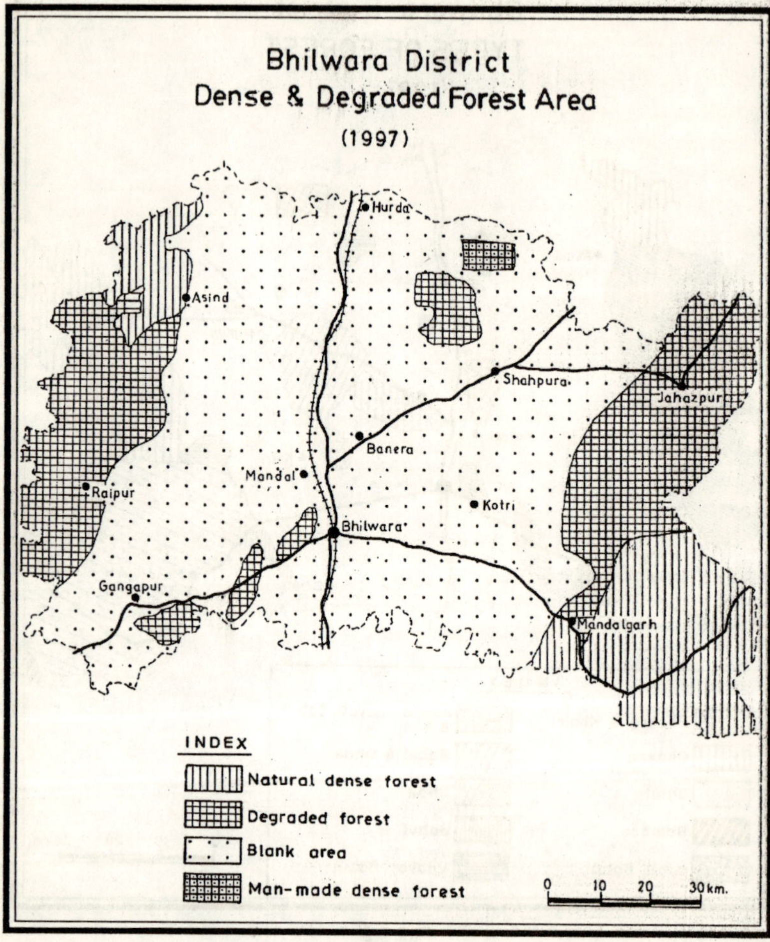

Fig. 5.3

forest types and biotic resources etc. may be considered as an important element of the eco-system on the whole. The vegetal cover, animal types and climatic conditions together form an eco-system of a region, where man and his behaviour play an important role in the development of sustainable environmental conditions. Man as a potent factor determines the conditions for the social forestry by utilizing natural resources around him. Thus the environmental impact may very well be visualized from the man's action and the extent of degradation of environment.

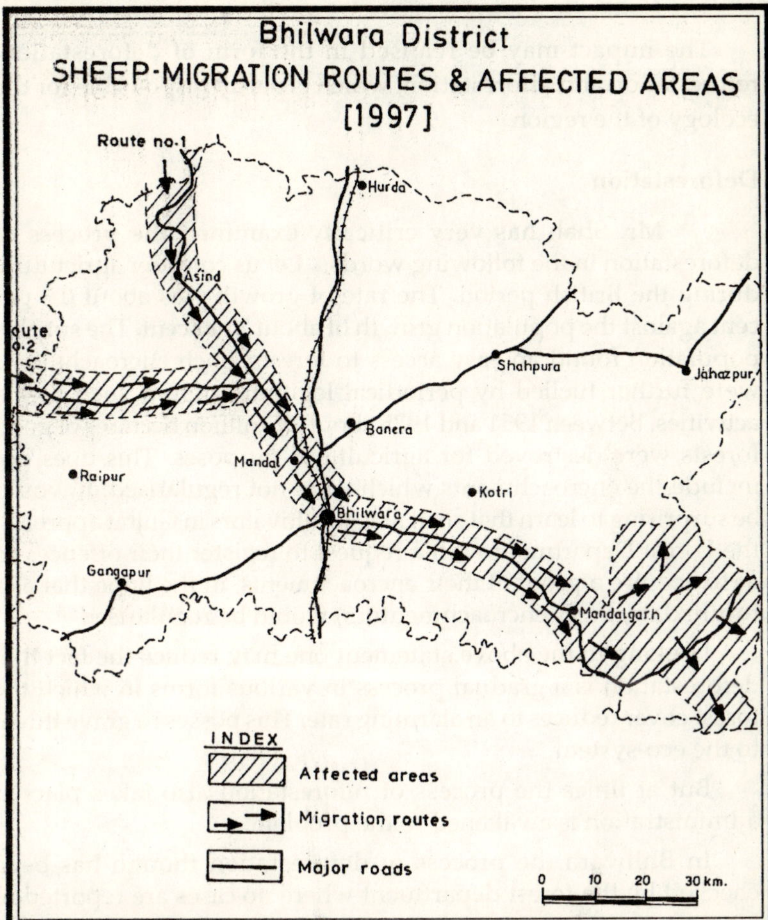

Fig. 5.4

Problem of Social Forestry

Social forestry is mainly a product of forest based resource or in other words fulfilment of human needs by forests. But the problem arises when the eco-existance comes to stand still or conflict, resulted in tense relationship between man and environment *e.g.* man cuts down trees beyond limit which adversely affect the rainfall and temperature conditions. This process is observed in a long chain reaction.

The impact may be realised in the form of deforestation, reforestation and afforestation, which is partly responsible for the ecology of the region.

Deforestation

Mr. Shah has very critically examined the process of deforestation in the following words, "Let us consider agriculture during the British period. The rate of growth was about 0.3 per cent against the population growth of about 2 per cent. The surplus population found an easy access to forests. Such encroachments were further fuelled by periodical legitimisation of such illegal activities. Between 1951 and 1975 about 2.5 million hectares of good forests were destroyed for agricultural purposes. This does not include the encroachments which were not regularised. It would be surprising to learn that such illegal cultivators in Gujrat approach the Forest Department with a request to register their offences so as to acquire a proof of their encroachments, in the hope that at a future date, they (encroachmentors) would be regularised"[5].

Looking to the above statement one may reduce the fact that deforestation is a gradual process in various forms in which the forest cover reduces to an alarming rate. This possess a grave threat to the eco-system.

But at times the process of reforestation also takes place if administration is awakened to the problem.

In Bhilwara the process of deforestation though has been checked by the forest department where no cases are reported of mass cutting down of trees by people or community for fuel purposes, but evidently it is also noted that down troden people of the society by and large are engaged in destroying forests in some way or the other. At the same time it is also noted that the deforestation process is going on without notice by the over grazing of animals which can not be checked out.

Knowingly or unknowingly the large scale quarrying activity in the vicinity of the heart of forest areas is causing great harm to the plants and thus forest cover is being reduced by intoxic substances, chemicals and dust.

But on the contrary a traditional practice of grazing animals, particularly sheep and goats causing great harm to the plants and thus reducing the forest cover. It is evidently noted that in Bhilwara district the dense forests are located in the south eastern part. A traditional migratory activity of sheep is well established since long, which has permanent routes of migrations, without of any interruption. Route No. 1 is located at the north-western part of the district adjoining Badnor and Asind, the route number No. 2 opens near Mandal town and runs together towards uppermal, the dense forest areas (Fig. 5.4).

The seasonal migration of sheep numbering about 1.2 lac are reported to be fled in period of July to October last. This process appears to be a very small and negligible but in consequences is greater than human interference as they do not allow plants to regenerate on a large scale.

Reforestation and Afforestation

Reforestation is a process in forested areas of where trees have been removed either by man or nature. Thus it may be treated as human action or artificial regeneration in degraded forest land. In Bhilwara district, area under reforestation is not very much high as compared to the total forest area, but it is considered to be significant when analysed in a different perspective. The reforestation area atleast speaks of the fact that the area covered under reforestation was by and large formerly under tree cover. The fact is an indirect expression of deforestation. Therefore if the figures of the reforestation since 1983 onward are analysed we find very interesting cases (Table 5.1). In 1983 the reported area under forest land plantation was accounted for about 400 hectares or 46.2 per cent of the total area under plantation, whereas in 1985 it rose to 800 hectares or 80 per cent of the total area under plantation. Here the total area under plantation includes the land of afforestation also. Thus out of this land if the 80 per cent land falls under reforestation, then it has got a different meaning. Similarly in 1986 area under reforestation to the total area under plantation is recorded as high as 72.8 per cent. Similar trend is recorded again in 1993 where it comes to 71.8 per cent. That in all the previous years under the new scheme of plantation area under reforestation has been higher as compared to the plantation of new areas, which is called as the afforestation. Therefore it may be understood that forest by and large are being destroyed in a gradual process in fulfilment of the basic needs of the local people.

Table 5.1 : Afforestation and Reforestation in Bhilwara (1983-93)

	1983 to 1993 plantations				
Year	*Community land plantations (Afforestation)*		*Forest land plantation (Reforestation)*		*Total area under plantation*
	In ha.	*%age*	*In ha.*	*%age*	*In ha.*
1	2	3	4	5	6
1983	464.0	53.8	400	46.2	864.0
1984	267.0	25.5	780	74.5	1047.0
1985	200.0	20.0	800	80.0	1000.0
1986	320.0	27.2	1175	72.8	1495.0
1987	360.0	33.6	710	66.4	1070.0
1988	831.5	51.1	795	48.9	1626.5
1989	989.0	54.8	816	45.2	1805.0
1990	660.0	30.5	1500	69.5	2160.0
1991	610.0	31.5	1325	68.5	1935.0
1992	515.0	33.2	1035	66.8	1550.0
1993	230.0	28.2	585	71.8	815.0
Total	5446.5		9921		15367.5

Source : Compiled from D.F.O., Bhilwara.

On the other side the afforestation is a community land plantation which is being developed along with reforestation. The objectives of the afforestation are the same as to protect environment on one side and to fulfil basic needs of the people on the other. The area under community land plantation is recorded much less than the reforestation area. (Table No. 5.1). This speaks of the fact that the community land plantation scheme is yet to get momentum.

References

1. Forest Department, Government of Rajasthan, Jaipur. 1996, State Forestry Action Plan (1996-2016). P. 15.
2. Mathur, C.M. 1983, Unpublished Ph.D. Thesis.
3. Forest Department, Government of Rajasthan. 1996, The working plan for Bhilwara Forest Division.
4. Champion and Seth. 1968, General Silvi-culture for India. Govt. of India Publication, Delhi.
5. Shah, S.A. 1993, Forestry Activities are not an adequate answer to deforestation. Waste land News. Feb-April, P. 14, New Delhi.

Chapter-6

Community Land Plantations

Society and Plantation

As it has been mentioned earlier that the community land refers to the panchayat land, which has been brought under the plantation schemes. Sometimes it also includes the Government waste land lying without of any use. Infact the community land plantation schemes were launched by the government with the objectives of providing fuel, fodder and food in order to release the pressure on existing forest areas on one side and to protect environment on the other.

The basic idea behind the community land plantation was to develop forest canopy along with rural areas so that the further degradation of forests may be checked out. It was also a well stressing point that the gradual degradation of forest cover is deteriorating the environmental conditions day by day. So it needs care well in time. Therefore, the answer was sought in the form of community land plantations.

In the recent past the unprecedent rise in the population all over the world in general and state of Rajasthan in particular where phenomenal growth of population has been at the alarming rate (2.1 per cent per annum), has exerted more pressure on the natural resources. The consequences of this fact were realised by 1980s when environmental conditions become more precarious and hazardous, and hence threaten the very existence of life on the earth.

This fact stressed the need of adequate forest cover (33 per cent) to maintain the balance between the man and environment.

The various plantation schemes *e.g.* community land plantation, farm forestry *etc.* were introduced mainly to substitute the loss or degradation caused by man's action.

Social Needs

It is an historical fact that the human societies at the beginning were settled in areas of luxuriant forest and fertile tracts of hill side. These areas were fascinating the culturable land under the tree cover. Thus, the little canopy of forests provided the shelter to the ancient human groups. The shelter was in forms of forest products and in form of the thatched huts. Though there was a danger of falling prey of wild animals but at times hunt them for food. This was a common practice but it reflects the importance of forests and forest products.

Forests provided the basic needs like food, fuel, and fodder to man and animals since long. So far the basic needs are concerned some of the items may be considered important for the district as a whole which by and large depends on forests products. Wood may be considered as the formost important item in use by the people as firewood and commercial wood. Firewood makes an important point as a fuel for the poor people. The concentration of fuel wood in Bhilwara town alone can be make out from the table below (Table 6.1).

Table 6.1 : Consumption of wood and Bans-Bullies

S. No.	*Item*	*Consumption*	
		Daily	*Annual*
1.	Fire-wood	210 tonnes	76650 tonnes
2.	Commercial-wood	150 Cum.	54750 cum
3.	Firewood for Funerals	40 bodies × 1/2 ton.	7300 tonnes
4.	Bamboos	1500 Nos.	547500 Nos.
5.	Bullies (Wooden poles)	550 Nos.	200750 Nos.

Source : Survey conducted in Bhilwara city.

From the above table it is estimated that the consumption of fuel wood is about 83950 tonnes per annum. This includes the fuel wood used in the urban area of Bhilwara city. If we count the total urban households as per the 1991 census report, it comes about 62000 of which 16680 families have been provided the L.P.G. gas connections. That means about 45000 families are still depending on the fire wood collected from different sources. It may be noted that out of the 45000 families about 21000 families have been provided the Kerosene facility at times. Thus remaining about 24000 families or household exclusively depending on the fire-wood. This reflects that some way or the other these families are getting fuel-wood from the forest areas. If we count the rural house-holds which accounted for about 256000 are barley depending on the fuel wood as no gas facility has been extended to them. Thus rural population is also fetching the firewood from different sources. In other words we may conclude that 77 per cent of the fuel requirement is fulfilled by the forests where as 23 per cent requirement is made by gas and kerosene etc. All these facts clearly reveals that the largest section of the society in Bhilwara district is still depending upon the indigenous sources of fuel *i.e.* fire-wood, dung etc. This point makes an important aspects for the development of social forestry.

It is interesting to note that here the need of firewood for the funeral purposes alone, if estimated as per the monthly basis, (case studied in Bhilwara city alone), it comes about 600 tonnes firewood for the approximately 1200 dead bodies. This wood if counted for a year it comes about 7300 tonnes.

Similarly if we come to the commercial wood used for different purposes particularly house building, furniture etc. It comes about 54750 cum. in a year. If we include bamboos and wooden poles (Bansbullies) which accounted for 1500 and 550 in daily supply respectively, accounts 547500 and 200750 annually. This fact also reflects the consumption of the commercial wood as well as the needs.

So far the fruits are concerned, they too have significant role in the rural economy as many families are engaged in the collection of Timru, Anwala, Ber, Khajur, Baheda, Mahua, Jamun, Ker, Sangri etc. The economic importance of the different fruits may be analysed from the table below.

With a special reference to ker economy in Banera panchayat of Banera Panchayat Samiti in Bhilwara district the 'Ker' auction in 1996 was reported 3 lac rupees. It may be noted that in about 120 to 140 village panchayat are having natural growth of 'Ker'. Similarly Timru, Anwala, Khajur and Mahua occupying a good place in the overall economy of the rural people.

Table 6.2 : Estimated Fruit Production in The District.

S. No.	Name of Fruit	Annual Production approximately in tonnes	Market value in thousand Rs./Ton	Total cost of the produce in lacs.
1.	Timru	250	4.0	10.0
2.	Anwala	60	5.0	3.0
3.	Ber	4000	1.5	60.0
4.	Khajur	250	6.0	15.0
5.	Baheda	50	8.0	4.0
6.	Mahua	40	10.0	4.0
7.	Jamun	20	11.0	2.2
8.	Ker	750	6.0	45.0
9.	Sangari	100	10.0	10.0
	Total	5520		153.0

Source : Field survey in Bhilwara district.

Fodder constitutes an important role in the social forestry. So far the fodder alone is concerned which includes grasses, leaves and pods etc., it form a basis for the animal based economy. Because animals are primarily depending on the fodder grown in the forests. In the district the fodder areas are reported as about 80000 hectares on which grasses are grown in the forests. The total live-stock number is accounted about 18.6 lac cow units in the district. The yearly production of grasses is reported about 2.0 lac tonnes from forests where as it needs 33.4 lac tonnes in a year. Thus the forests are providing very meagre amount of grasses to the animals. Though the larger amount of fodder comes from the agricultural products which fulfils the needs.

It is interesting to note that the forest department has evaluated the average feed of the fodder assuming cow as one unit. On this basis the different category of animals as per the table 6.3, the fodder consumption is 3339421. 2 tonnes.

Community Land Development

The community land development, here, as it is described in the second chapter, meaning by land owned by the panchayat samities or in a restricted sense owned by religious group of people in village sites, is considered important from the community land plantation development. In general usage this type of land is generally called as "Charnot". Therefore the social forestry development programme is carried on with the consent of the village panchayat or panchayat samities. Social forestry hence, cannot be developed unless and until the village panchayats are agree to do so. Thus panchayat samities play an important role in development of social forestry schemes.

Table 6.3 : Fodder Requirement in the District

S. No.	*Category of animals*	*Number of animals*	*Conversion factor for cow unit*	*No. of cow units.*	*Annual require-ment of grass in tonnes/ unit*	*Total requirement in tonnes*
1.	Cow	777029	1.00	777029	1.8	1398652.2
2.	Bufallo	333242	2.00	666484	1.8	1199671.2
3.	Camel	13055	4.00	52220	1.8	93996.0
4.	Goat	680132	0.25	170033	1.8	306059.4
5.	Sheep	679128	0.25	169782	1.8	305607.6
6.	Others	19686	1.00	19686	1.8	35434.8
	Total	2502272		1855234		3338969.4

Source. District animal husbandry office Bhilwara.

In this respect Bhilwara district is divided into eleven panchayat samities, they are Ashind, Banera, Hurda, Jahazpur, Kotri, Mandal, Mandalgarh, Raipur, Sahada, Shahpura and Suwana. The biggest panchayat samiti in respect of area is Mandalgarh, occupying about 148.8 thousand hectares area whereas the smallest one is Raipur with 51.9 thousand hectares area. All these panchayat samities possess varying size of community land[1], which is important from the social forestry point of view. The community land under different Panchayat samities reflects the level of community land development schemes against the geographical area (Table 6.4).

It is interesting to note that the average area under community land comes about 11.2 percent of the total geographical area. The Banera Panchayat samiti holds 14.6 percent community land of the geographical area, which is highest in the district while the second highest is of Kotri (12.6 percent) and equally Sahada (12.5 percent). Rest of the panchayat samities (Tehsil) are occupying 8 to 11 percent community land of the geographical area. These facts show that the community land area under each of the panchayat samities having very meagre area or about 10 percent, of the geographical area.

Table 6.4 : Panchayat Samiti Wise Community Land Plantation (1983-1993)

S. No.	*Name of Panchayat Samiti*	*Total Geographical area (in 000 ha.)*	*Community land (in 000 ha.)*	*Community land %age to the total area*	*Plantation done 1983-1993 (in 000 ha.)*	*Plantation done 1983-1993 %age to community land*
1.	Asind	113.7	13.2	11.6	0.530	4.6
2.	Banera	68.0	9.7	14.6	0.312	3.2
3.	Hurda	61.7	6.6	10.7	0.269	4.6
4.	Jahajpur	108.4	10.5	9.7	0.090	0.6
5.	Kotri	92.8	11.7	12.6	0.285	2.4
6.	Mandal	120.6	11.4	9.5	0.384	3.5
7.	Mandalgarh	148.8	12.3	8.3	0.629	5.1
8.	Raipur	51.9	6.0	11.6	0.382	6.6
9.	Sahada	65.0	8.4	12.5	0.758	5.0
10.	Shahpura	112.8	12.8	11.5	0.851	6.6
11.	Suwana	96.5	10..6	10.6	0.954	9.0
	Total	1040.2	113.2		5.444	
	Average			11.2%		4.6%

Source : District Statistics Report 1994, & District Forest Office, Bhilwara.

If we look to the another aspects of the community land plantation, it provides a different perspective of the social forestry in the district. The community land already covered under plantation speaks a different story. The area covered under plantation in different panchayat samities, comes to 4.6 percent of the community land as an average (Table 6.4). The highest area under plantation cover is of Suwana panchayat samiti *i.e.* 9.0 percent of the total community land, whereas the lowest is 0.6 percent of the Jahazpur panchayat samiti. It is interesting to see that while both these panchayat samities are having almost equal community land *i.e.* 10.6 and 10.5 thousand hectares respectively, but marked with a wide gap of existing plantations. This speaks inequality of the plan development.

From the social forestry point of view it may be borne out that Jahazpur occupying larger area under forests tree cover as compared to suwana panchayat samiti and dominated by meena community. Suwana panchayat samiti is located quite near to the city area and hence easily vulnerable by the local people for fuel wood and other needs. Therefore the forest cover around city is more degraded than the remote areas.

Plantation Schemes

Plantation Methods and Technology

In social forestry one has to adopt a particular line of plantation method and technology. It needs careful evaluation of land classification and the selection of various tree species. The former point deals with the land terrain site, soils etc. Whereas the latter refers to the tree species compatible in the existing environmental conditions. The local needs particularly for fuel and fodder, are to be considered in the light of the above statements.

It is also a well born fact that social forestry on community land or land owned by panchayats is occupying very less area, which is insignificant to the total geographical area, but it holds a key of the social forestry.

The various methods employed in plantation schemes are for varied purposes like Fuelwood plantation, Timber plantation, Silvipastoral plantation, Rehabilitation of barren hills (R.B.H.) or (R.D.R.) Rehabilitation degraded forests and plantation along with the agricultural fields etc., etc. Sometimes looking to the situation plantation is done alongwith the canals and roadsides. Therefore, the plantation schemes depends on the need of the people as well as the situation.

In the present context the social forestry is being emphasized mainly for the fuel and fodder purposes in order to save forests cover. In Bhilwara district the community land plantation schemes are enhanced with the all perspectives as mentioned above but with a limited scale.

(a) Fuelwood Plantation (F.W.P.)

In every village panchayat barren pasture land is used for fuelwood, grasses and timber wood. Therefore this type of land is used for community plantation. The particular method employed for tree plantation for fuel purposes is to conserve water and to check soil erosion. For this purpose contour trenches or V-ditches in the length of about 400 Rmt. are drawn. Grasses are developed alongwith the tree avenue. For this purpose particular tree-species which provide the fuel, Babul (Acacia nilotica), Israili Babul (Acacia tortlis), Siras (Albizzia lebbek), Neem (Azadirachta indica), Vilayti Babul (Prosopis juliflora), Ber (Zizyphus jujuba), Khejari (Prospis spicigera), Ronj (Acacia leucophloea) etc. are grown. For sapling plantation 1100 pits of 45×45×45 cm. are dug.

(b) Silvi-Pastoral Plantation (S.P.P.)

Under this plantation scheme grasses or fodder for the animals in dry part of the state is raised. It is estimated that about 1095 hectare land is brought under collective farming for the S.P.P., on every hectare 200 plants are planted. The particular species of plants is favoured by Babul (Ac. nilotica), Khejari (Prosopis spicigera), Neem (Az. Indica), Ardu (Aillanthus excelsa), Ber (Zizyphus jujuba) etc. Under these plants grasses are grown for animals as suitable fodder.

(c) Rehabilitation of Barren Hills (R.B.H.)

Under this method the rehabilitation of barren hills is done in a particular manner.

(*i*) 1100 saplings are planted in pits of 45 × 45 × 45 cms size,

(*ii*) Contour or V-ditch are made of 400 metres length, per hectare,

(*iii*) Contour furrows are made of 50 metres per hectare.

The plants favoured under this scheme are Khair (Acacia catechu), Kumta (Acacia senegal), Churel (Holoptelia Integrifolia), Dhak (Butea monosperma), Jungle Jalebi (Pithocolobium dulce), Bans (Dendrocalamus strictus), Neem (Az. indica), Vilayati Babul (Pro. juliflora) etc.[2]

(d) Rehabilitation of Degraded Forests (R.D.F.)

In different parts where forests are degraded planting is done to check further degradation. For this purpose two different models are applied *i.e.* model I and Model II. Model-I is applied in areas where intensive degradation have taken place (Crown density 10 to 40 percent), under this method 300 Rmt. contour trenches are made with 50 Rmt. contour furrows and 700 pits of 45 × 45 × 45 cm. per hectare.

The model-II is applied in area of moderate degradation where crown density exceeds over 40 percent under this method gap plantation is done with 200 plants in a hectare. It is generally done with a random sampling. For water conservation staggered contour trenches are made of 400 meters length per hectare. The plant species favoured under this scheme are same as in R.B.H.

Looking to the above methods and techniques the development of social forestry in Bhilwara district may be judged from the table[3] (Table 6.5).

The development of social forestry on community land during the last 10 years has been in a different forms. The area covered in different schemes during the period 1983 to 1993, fuel wood is recorded as high as 781 ha. land in Shahpura Panchayat Samiti, the lowest land under the scheme was marked in Mandalgarh *i.e.* 14 hectares. This speaks of the range of development during the last one decade. Though other panchayat samities like Asind, Kotri, Mandal, Raipur, Sahada, Suwana, etc. are fairly well in fuel wood plantation.

Similarly silvi-pastoral plantation is marked highest in Mandalgarh by 600 hectares, speaks of the demand of fodder for the animals. It is interesting to see that the demand of fuel wood is negligible, whereas fodder is higher, it is marked with a special obligation to the area on the name of development.

It is also interesting to see that rehabilitation of degraded forest is marked in Asind Panchayat Samiti only by 60 hectares of land in the district during the last 10 years. It speaks that no other parts are reported to be damaged during the period. But it is because of the fact that the forest land development is under the direct control of the forest department. Hence social forestry for the rehabilitation purposes of degraded forests is not reported for this purpose.

Thus on the whole it may be borned out that the main objective of the study is concentrated on the social forestry with the object that local needs are met with the adequate supply of fuel and fodder, leaving aside the other activities. Therefore the above table justifies the very purpose of the study, that whatever community land is available near by the town is being used for the purposes.

Table 6.5 : Community Land Plantation under Different Schemes (in Hectares) (1983-1993)

S. No.	Name of Panchayat Samiti	Fuel wood	Silvi pastoral	Timber	R. D. F.	R. B. H.	Shelter belt	Total	Total community land	Percentage to community land
1.	Asind	410	45	15	60	—	—	530.0	13529	4.6
2.	Banera	257	—	—	—	—	55.7	312.7	9585	3.2
3.	Hurda	131	65	15	—	—	58.3	269.3	6706	4.6
4.	Jahajpur	90	—	—	—	—	—	90.0	10566	0.6
5.	Kotri	240	15	30	—	—	—	285.0	11641	2.4
6.	Mandal	313	50	—	—	—	21.0	384.0	17469	3.5
7.	Mandalgarh	14	600	—	—	—	15.0	629.0	12426	5.1
8.	Raipur	310	45	—	—	25	2.0	382.0	6047	6.6
9.	Sahada	540	130	4	—	50	34.7	758.7	8365	5.0
10.	Shahpura	781	70	—	—	—	—	851.0	12824	6.6
11.	Suwana	752	75	90	—	—	37.8	954.8	10654	9.0
	Total	3838	1095	154			224.5	5446.5	119812	

Source : D. F. O. Bhilwara 1996.

Case Studies

For the purpose of the above stated explanations for the social forestry in general and different schemes in particular some case studies are provided here. These case studies have been made in order to provide full justification with a detailed account of the work done in different parts of the district *i.e.* community land area of panchayat samities.

(I) Community Fuel Wood Lot Rajpura-1987 (30 ha.)

General Information. The community wood lot was raised in the year 1987, on community land in village Rajpura of Sahapura Tehsil, in Bhilwara district. The village is situated at a distance of 79 Kms. from Bhilwara and 28 Kms. from Shahpura on Phulia-Dhanop road. The total community land is 171 ha. in the village out of which planting has been taken up over 30 ha. in 1987. As the villagers and Panchayat saw the success of plantation gave more land for F.W.P. The total community land which has been brought under afforestation by the forest deptt. is 102 ha.

The village panchayat is the legal owner of the community land. The village panchayat entrusted the task of raising the woodlot to the forest department in the year 1987, in pursuance of the government order lay-down that the forest department shall take up afforestation and pasture development work, and maintain the plantation area for 5 years and after wards hand over the area back to the panchayat, for protection and management, as per the management plan of the area prepared by the forest department.

The harvesting of trees and reforestation of the area will be taken up in the joint supervision of the village panchayat and forest department. The revenue from the wood and other produce shall be shared on 50 : 50 basis between the village panchayat and forest department.

Prior to the raising of fuel wood plantation, the area was used for grazing of cattle.

Establishment and Maintenance. The village wood lot over the community land was raised in the year 1987 under the scheme of village fuel wood lot in National Social Forestry Programme. The area was surveyed and fenced with trench of 90 × 90 Cm. in cross section.

The soil is saline with a Kankar pan at places, at 0.30 to 0.60 Mts. depth. In some patches the Kankar pan is exposed.

Planting pits were dug, at a spacing of 2.5 × 2 metres and planting of Acacia tortlis, Acacia nilotica and Zizuphus jujuba was taken up with the onset of monsoon.

The number of planted saplings is as follows :

Acacia tortlis	28,000
Acacia nilotica	21,000
Zizuphus jujuba	1,400
Total	50,400

Line sowing of presopis juliflora was taken up on the trench mound to raise a live hedge. The area was looked after by a cattleguard, employee of forest department.

Present Status of the Fuel Wood Lot. The survival of saplings is 85 percent but about 70 percent plants have put up satisfactory growth because the area is overlain with Kanker pan. Acacia nilotica and Acacia tortlis has developed into small to medium size trees in fairly deep soil but remained stunted in patches of lime Kankar pan. The Ziziphus jujuba has developed into small size trees and producing fair amount of fruits. The girth of the trees varies from 10 cm. to 30 cm. And the height from 3 to 6 meter. The average girth is 18 cm. and average crop height is 5 meter.

The survival of Acacia nilotica saplings is about 80 percent. The average girth of Babul plants is 25 cms. And average height 5 meter.

Table 6.6 : Survival and Growth in Rajpura Village Wood Lot

Tree species	*No. planted*	*Estd. survival % age*	*Estd. height*	*Estd. girth at collar*
Acasia tortlis	28000	75	4 m.	20 cm.
Acasia nilotica	21000	80	5 m.	25 cm.
Ziziphus jujuba	1400	80	4 m.	15 cm.

Spacing. The seedlings were planted at a spacement of 2.5 × 2.0 mt.

The growth of planted saplings shows variation due to soil quality, varying soil depth and presence of lime Kankar pan.

Objects of Raising Community Fuel Wood Lot. The objects of raising the community fuel wood lot are as follows :

(*i*) To meet the fire wood and fodder requirement of the community.

(*ii*) To utilize the fallow land of the community.

(*iii*) To provide employment opportunities to the rural landless, people.

(*iv*) To create awakening among the people, to raise trees on the marginal land of their fields and homesteads.

(*v*) To create an asset for the village panchayat to generate income.

(*vi*) To minimize soil erosion, provide shade for the cattle and ameliorate climatic conditions.

Description of Harvesting Method. The produce will be harvested by the villagers themselves, initially by prunning the shoots after 5th year onwards.

The prunning will be done in the presence of a forest guard/ cattle-guard and supervised by the Range Officer. The suppressed plants will be removed with prunning operation. The marking of suppressed plants and prunning height of the plants will be done on 1 ha. land in advance to serve as demonstration area an if it is felt that the operation can be carried out on remaining area without marking, it will be done in the presence of a forest official. The fire wood available would be about 70 tonnes.

Thinning. Thinning operation will be done in 8th year by removing the dead and suppressed plants. Then the plants in lime Kankar which have not put up good growth (height below the 3 meters) will be removed to provide the space for growth of the remaining plants. The estimated yield by thinning would be approximately 196 tonnes.

Final Harvest. The final harvest would be taken up in the year 2005. The rotational age of the tree is calculated to be 18 years for this area. When the plants starts to die or retards its growth.

The final fellings will be done by the contractor offering the highest bid in open auction after *vide* publicity and the felled material will be stocked in 3 lots; (*i*) Timber billets of more then 50 cm. girth over bark, (*ii*) Fuelwood billets of more than 20 cm. girth over bark and (*iii*) fuelwood less than 20 cms. girth over bark.

The firewood of less than 20 cms. girth over bark will be given free or at concessional rates to the villagers. The concessional rates will be decided by the village forests management committee or if the committee decides to sell the entire wood and utilize the proportionate income on village development or welfare activities. The produce will be sold in open auction in the presence of social forest officer.

There are 101 house-holds in the village (Total population 602) and the produce comprising of fire wood of less then 30 cms. girth over bark shall be distributed equally free among the villagers on the advice of the members of village forest protection and management committee in accordance with the Rajasthan Government order dated 26th April, 1991. The fuelwood available for each family would be 0.35 ton.

Protection and Maintenance. The cattle guard and forest guard incharge of other adjoining plantation will also keep watch over the area. No expenditure is being incurred on the protection of the area.

The village forest protection and management committee shall decide the penalty or fine for illcit felling of free by their own villagers and others. They may also decide the levying of fee for head loads of firewood or the concessional rates for sale of firewood in consultation with their own village people in the Gram Sabha, in the presence of the local forester who is also the member secretary of the working committee.

Regeneration Plant. After the final harvest the area will be ploughed with a bullocks drawn country plough or with the disc plough, drawn by a tractor, as the village forest protection committee decides along the contour and sown the seeds of, Dhaman grass, legumes like Jhari ber will be planted up for the next rotation from the 50 percent income of the village panchayat. The social forestry wing will provide additional funds for creation of community woodlots.

Accounts and use of funds. The sale proceeds from larger size firewood and timber shall be shared on 50 : 50 basis between the village panchayat and forest department. The share of the panchayat shall be deposited into the accounts of village panchayat/community woodlot management committee to be operated jointly by the forest Range Officer and Secretary village panchayat/cashier of the management committee. 25 per cent of the sale proceeds shall be spent on reseeding with grasses legumes, shrubs (Jhari ber).

Analysis. The production of the firewood is likely to meet 30 percent requirement of fuelwood (Table 6.7). There are also other plantations in the village and hence will meet the requirement of fuel wood. It is calculated that about 90 per cent demand would be met out from these plantations. The villagers are of the view that they would meet the fuel wood requirement from other trees growing over community lands, revenue wastelands, roadsides, farm yards, and marginal land of fields. There is need for creating tree consciousness among the people and to motivate them for planting trees on the marginal land of their fields. It will also help in saving the cowdung (to be utilized as organic manure), part of which is burnt for cooking the food.

II. Community Wood Lot Raila

(Banera panchayat samiti)

1. Name : Raila Plantation.
2. Area : 20 ha.
3. Year of planting : 1990.
4. Status of land : Panchayat (Community) land.
5. Soil type : Loamy soil with Kankar pan at 30 to 50 cm. depth.
6. Type of fencing : Livehedge thor fencing was done, inside thor fencing A. nilotica and prosopis juliflora was sown to assist the thor fencing.
7. Soil working : 22,000 pits of size-45 × 45 × 45 mt. were dug, at a spacing of – 3 × 3 mt.
8. Seed sowing : (*i*) Seeds of A. nilotica and prosopis was sown on inside the thor fencing.
9. Species planted : (*i*) A. nilotica 17940
   (*ii*) A. leucophloea 3200
   (*iii*) Ziziphus jujuba 860

Table 6.7 : Yield Table (Species wise)

Species	*No. Planted*	*Survival estimated (70%)*	*Estimated out turn*	*Unit*	*In 8th year*	*9th yr.*	*10th yr.*	*11th yr.*	*12th yr.*	*13 yr. onwards*
Acacia	21,000	14,700	In 8th year	M.T.	30	30	30	30	–	–
nilot-			0.002 ton Fuelwood/tree							
ica			In 12th year	M.T.	—	—	—	—	44	44
			0.003 ton Fuelwood/tree							
			0.001 ton Pods/tree	M.T.	15	15	15	15	30	30
Acacia	28,000	19,600	In 8th year	M.T.	39	39	39	39	—	—
tortlis			0.002 ton Fuelwood/tree							
			In 12th year	M.T.	—	—	—	—	59	59
			0.003 ton Fuelwood/tree							
			0.001 ton Pods/tree	M.T	20	20	20	20	40	40
Ziziphus	1,400	980	In 8th year	MT.	1	1	1	1	—	—
Jujuba			0.001 ton Fuelwood/tree							
			In 12th year	M.T.	—	—	—	—	0.75	0.75
			0.002 ton Fuelwood/tree							
			0.002 ton Fruit/tree	M.T.	2	2	2	2	2	2
			0.002 ton Fodder/tree	M.T.	2	2	2	2	2	2

Abstract : Pods, fruit and fodder given free to the villagers for the welfare of the society.

10. General condition	:	The plants have attained the average height from 2.5 to 3 mt.
11. Yield results	:	After 4th year onwards lopping of the plants will be done and this intermediate yield will be distributed among villagers free of cost. On an average 1 Kg. of fuel wood per plant will be made available to the villagers. Approximately 20 tonnes of fuelwood will be made available to the villagers.
		Thinnings and final harvesting of the plantation would be carried out according to the silvicultural management plan.
		General information, objective of plantation and benefit sharing is as dealt in Rajpura fuelwood lot.

III. Community Wood Lot Berra

(Fuel wood plantation Benera panchayat samiti)

1. Name	:	Berra
2. Area	:	30 Ha.
3. Year of planting	:	1990.
4. Status of land	:	Panchayat (Community) land.
5. Soil type	:	Clay, Loamy soil with Kankar pan at 30 to 50 cm. depth.
6. Type of fencing	:	Ditch fencing of {(1.5 + 0.9) × 1.2}/2m cross-section. It is strengthen by Prosopis juliflora sowing on the mound.
7. Soil working	:	1. 1704 Box trenches of size 2 × 0.45 × 0.45 m and 2528 trenches of size 1.5 × 0.45 × 0.45 were dug, for moisture conservation.
		2. 38100 pits of size 0.45 × 0.45 × 0.45 m were dug, for planting.

8. Seed sowing	:	(*i*) Seeds of Dhaman grass was sown on box trenches.	
		(*ii*) Sowing of tree species (P. juliflora) on the mound of ditch fencing and on box trenches was done.	
9. Species planted	:	(*i*) A. nilotica	30341
		(*ii*) A. leucophloea	6629
		(*iii*) Ziziphus jujuba	1130
		Total	38100
10. General condition	:	Now plants have attained the average height from 2.5 to 3 mt.	
11. Benefits	:	Free cutting of grass was allowed by head loads to local villagers. In the year 1990 and 1991 grass approximately 12 and 20 tonnes respectively was distributed free of cost among villagers.	

IV. Community Wood Lot Phulia

(Fuel wood plantation Shahpura P. S.)

1. Name	:	Phulia
2. Area	:	30 ha.
3. Year of planting	:	1987
4. Status of land	:	Panchayat (Community) Land.
5. Soil type	:	Clay-alkaline with lime Kankar lane, at places it is exposed at the surface.
6. Type of fencing	:	Ditch fencing of 60 × 60 sq. cm. cross section was dug.
7. Soil working	:	45,000 pits of 50 cub. cm. at a distance of 2.5 mt. × 2 mts. were dug.
8. Seed sowing	:	On the outer side of the fencing prosopis juliflora seeds and on the inner side of the fencing Acacia nilotica seeds were sown.

9. Species planted	:	The saplings were planted as under :	
		1. Acacia nilotica	21,000
		2. Acacia tortlis	23,000
		3. Zizyphus jujuba	1,000
10. General condition	:	Average height of the plants is 5 meter. The survival of the plants is about 65 percent the area is very well developed and soil erosion is also checked.	
11. Benefits	:	Free grass by head loads was allowed to the villagers. The villagers are also getting the pods of Acacia nilotica, Acacia tortlis on an average 1 kg. pods/tree *i.e.* 30 tonnes. In this way the villagers are fetting fodder of about Rs. 20,000 in addition to the free grass for their cattle. Hence the plantation has achieved its objectives.	

V. Kanwaliyas Timber Plantation-1989 (15 ha)

General Information about Plantation : This community land plantation was raised in 1989 on panchayat land of village Kanwaliyas, tehsil Hurda, of Bhilwara District. The village is situated at a distance of 20 km. from Gulabpura on Ajmer Bhilwara road (S.H. –4). Total area of Gochar land is 175 hectare out of which planting has been taken up over 15 hectare in 1989. Plantation is situated at a distance of 2 km. from Kanwaliyas village in the eastern direction.

The village panchayat entrusted the task of raising the woodlot to the forest department in the year 1989 in pursuance of the Government Order No. F. 17 (II) Rev/8/82/dated 5th June 1982. The government order lays down that the forest department shall take up afforestation and pasture development works, protect and maintain the area for 5 years and then after hand over the area back to the panchayat for protection and management as per the management plan of the area prepared by the forest department.

The harvesting of trees and reforestation of the area will be taken up in the supervision of the village panchayat and forest department.

The revenue from the plantation shall be shared on 50 : 50 basis between the panchayat and forest department. Prior to the raising of woodlot the area was used for grazing of cattle by the villagers.

Establishment and Maintenance. The panchayat land plantation was raised in the year 1989 under the scheme of N.S.F.P. in the timber model. The area was survyed and fenced with a ditch fence, $\{(1.5+.9)\times 1.2\}/2$ mt. cross section to raise live hedge fencing sowing of prosopis juliflora, A. nilotica, parkin sonia and kumta etc. was done.

The soil is loamy with a Kankar pan at 40 to 60 cm. depth.

Planting was done at a spacing of 3 × 3 mt. Box trenches of size 125 × .45 × .45 were also dug at the spacing of 3 × 3 mt.

The number of saplings planted are as follow :

1. Shisham	(Dalberzia Sissoo)	800
2. Siris	(Albizzia lebbek)	1,700
3. Ber	(Zizyphus jujuba)	4,000
4. Babool	(Acacia nilotica)	11,000
	Total	17,500

According to model 16,500 plants should be planted but due to availability of space 1,000 additional plants were planted.

Dhaman grass (Cencrus ciliaris) seeds were sown on the mound of the trenches.

Present status of the wood lot. The survival percentage of plants in the plantation is above 90 percent but only 75 percent seedlings have established up well with satisfactory growth.

The average height and girth of the trees is as given below (Table 6.8).

Table 6.8 : Species wise survival and Growth

S. No.	*Species*	*No. of plants*	*Survival percentage*	*Average height*	*Average girth*
1.	Sisham	800	85	2.7 mt.	23 cm.
2.	Siris	1,700	85	2.5 mt.	20 cm.
3.	Ber	4,000	90	3.0 mt.	23 cm.
4.	Babul	11,000	90	2.4 mt.	20 cm.

The growth of planted saplings indicated remarkable growth due to soil depth.

Objectives of raising community wood lot.

The objectives of raising the community woodlot are as follow :

1. To meet the Timber and firewood requirement of the community.
2. To meet the fodder requirement of the community
3. To utilize the fellow land of the community.
4. To minimise soil erosion, provide shade for the cattle and to ameliorate climatic conditions.
5. To provide employment opportunities to the rural landless poor.
6. To create an asset for the village panchayat.
7. To create awakenning among the people, to raise trees on the marginal land of their own field.

Harvesting Method. The produce will be harvested by the villagers themselves initially by singling of shoots in the Ber plant in the 4 year. Singling will be done in the presence of forest guard/ cattle guard and supervised by the Range officer. Singling will be done every alternate year up to 10th years. The produce available by singling will be distributed free of cost to villagers. Grass will also be supplied free of cost.

The final harvest would be taken up in 2008 when the singled shoots have attained a diameter of more than 30 cm.

However decision will be taken to provide proper space and light to plants so that they can attain a good height and diameter.

Thinning will be done as follow (Table 6.9).

Table 6.9 : Thinning Plan

S. No.	*Species*	*I Thinning*	*II Thinning*
1.	Sisham	VII Year	X Year
2.	Siris	VIII Year	XI Year
3.	Ber	V Year	VII Year
4.	Babul	VI Year	VIII Year

Estimating 80 percent survival of trees at the time of final harvesting nearly 50 trees per hectare will be retained as seed bearers. The number of trees available for final harvesting will be as follows (Table 6.10).

Table 6.10 : Trees available for final harvest

S. No.	Species	No. of saplings planted	Survival 80 %	After II thinning	Seed bearer trees	Net available trees
1.	Sisham	800	640	160	35	125
2.	Siris	1,700	1,360	340	75	265
3.	Ber	4,000	3,200	800	175	625
4.	Babul	11,000	8,800	2,200	465	1,835
	Total	17,500	14,000	3,500	750	2,850

The trees will be felled at a height of not more than 15 cms. from ground level.

The coppice regeneration from final felling will again be subjected to singling of shoots after 2 years and II rotation felling of singled shoots would take place in 2023.

The final felling up will be done by the contractor offering the highest bid in open auction after wide publicity and felled material will be stacked in 3 lots :

(*i*) Fuelwood billets of more than 20 cm. girth over bark.

(*ii*) Fuelwood less than 20 cm. girth over bark.

(*iii*) Timber.

The firewood of less than 20 cm. girth over bark will be given free of cost. The produce will be sold in open auction in the presence of forest officer.

There are 439 household in the village (total population 2429) and the produce comprising of firewood of less than 20 cms. girth over bark shall be distributed equally free among the villagers.

Regeneration Plan. After the final harvest the area will be ploughed along the contour and sown with seeds of Dhaman grass, legumes and Jhari ber and protected from grazing and browsing. Additional area would be taken up from the 50 percent income of the village panchayat.

Accounts and use of funds. The sale proceeds from larger size firewood of more than 20 cms. girth over bark shall be divided on

50 : 50 basis between the village Panchayat and Forest Department. The share of the panchayat shall be deposited into the bank account and to be operated jointly by the forest Range officer and secretary village Panchayat.

VI. Timber Plantation Erans-1992 (15 ha.)

(Panchayat samiti Asind)

1. Name	: Erans	
2. Area	: 15 ha.	
3. year	: 1992	
4. Status of land	: Panchayat (Community) land	
5. Soil type	: Loamy soil with Kanker pan at 30 to 50 cm. depth	
6. Type of fencing	: Ditch fencing of size {(1.5 + 0.9) × 1.2}/2 mt. Cross-section. It was strengthen by prosopis juliflora, Ac. nilotica and Ac. Sengal sowing.	
7. Species planted	: 1. Babul	16,000
	2. Shisham	200
	3. Neem	50
	4. Ber	150
	5. Miscellaneous	100
	Total	16,500
8. Grass sowing	: Sowing of Dhaman grass was also carried out.	
9. General condition	: Average height of the plants is 4 mt. and survival is about 85 percent. The general information, objective and management is same as of Kanwaliyas timber plantation.	

From the above case studies the concurrence may be drawn that the social forestry in Bhilwara district has a wider impact on surrounding land areas and the resources available in the vicinity. In other words it may be said that with a gradual expansion of urban areas as well as the steady population growth compelling natural resources to banish. This needs a counter balance to maintain environment as well as to meet out the local needs of the people, mainly depending on the sorrounding areas. Failing of which may result the further degradation of environmental conditions. Therefore the community land development schemes and plantation may not be in a wider perspective, but in a limited way fulfil the nature's balance.

References

1. Directorate of Economics and Stastics, Rajasthan, Jaipur. 1994, District Statistical Report of Bhilwara.
2. Department of Forests, Rajasthan, Jaipur. 1995, Forestry Development Project, Series Ist.
3. District Forest Office, Bhilwara. 1996, Community land plantation, under different schemes from 1983 to 1993.

Chapter-7

People's Participation and Its Impact

In the development of social forestry people's participation is the key to the success of the scheme. It is the population in general and local people in particular to play an important role in shaping the development of social forestry. Because it is an inter dependence of forest and the needs of the people in vicinity of rural settlements. Therefore, in success of the social forestry schemes people's participation is a key factor. Besides that some governmental and semi governmental agencies are involved for the purpose. But looking to the importance of the social forestry schemes, some private agencies are also involved and attracted for the work particularly at the time when environmental problem crop-up.

Some international agencies are also invited for the development of forests and their allied activities. International agencies like World Bank (W.B), International Monetary Fund (I. M. F.) and Overseas Economic Co-operation Fund (O. E. C. F.) of Japan etc. are worth mentioning here. They are providing funds and working in collaboration with India. These agencies are providing huge funds under various schemes in particular. The state government's role in this process is to provide the infrastructural facilities. In spite of all these things the people's role is considered greater in handling the situation at the particular site. Hence the role of panchayats and panchayat samities are no way less significant. The village panchayat controls the village people and hence social forestry development is accelerated.

The success of the social forestry depends on the two levels i.e.

(*i*) National Waste Land Development Programme (N. W. D. P.) and,

(*ii*) Joint Forest Management (J. F. M.)[1].

It is enhanced that in the development of social forestry the people's participation is levelled or controlled in three stages which are shown in the following forms.[2]

Table 7.1 : Stages of Social Forestry

S. N.	*Form of management*	*Control of land*		*Forest department*
		Private	*Community*	
1.	Individual household	Farm forestry	Tree tenure	Tree tenure
2.	Community/ F. D.		Community land plantation	Community land plantation
3.	Forest Department (F. D.)		Community land plantation	Waste land plantation on Govt. land.

It is evidently known that land is controlled under private, community or panchayat and forest department, on private land where individual house-hold plays an important role. Farm forestry is developed but on the community waste lands whereas social forestry is developed under the control of forest department.

Joint Forest Management (J. F. M.) is being done right from the beginning in various parts of the state. But its true form and status was realized since the development of social forestry. Mr. R. P. Kapoor and his associates have rightly observed the J.F.M. in the Rajasthan State have highlighted some of the important aspects[3]. Under the J. F. M. "Janta Padocs"—a scheme of people's participation was involved, specially in the district of Bhilwara with the objective of protecting forest lands. Under this scheme the forest land is protected by erecting a boundary wall by the villagers. Within this boundary wall fodder for the animals is developed alongwith the fuelwood. Thus it is a scheme of direct involvement of the local people.

In Bhilwara district Mandalgarh Panchayat Samitee is a unique case in the whole district which has adopted, the "Janta Padoc Scheme" to protect their land from the invading animals more particularly sheep and goats from out side. It is interesting to mention that the migration of small animals is enrouted through uppermal area, (Fig. 5.4 sheep migration).

These animals (sheep and goats) stay here for about 4 months *i.e.* July to October. It is interesting that this is the high time for the germination of seeds. Therefore it is necessary to protect these plants from the animals. Hence the 'Janta Padocs' are the answer to the problem. In Mandalgarh Panchayat Samitee the area covered under the Janta Padocs is given in Fig. 7.1 and Table 7.2.

Table 7.2 : Janta Padocs

S. No.	*Name of village*	*Name of Forest Block*	*Panchayat Samitee*	*Area in Hect.*
1.	Amarniwas	Raila Kharla	Mandalgarh	80.0
2.	Sunthi	Bhopatpura	—do—	50.0
3.	Laxmi Khera	—do—	—do—	154.0
4.	Kama	Thumasa	—do—	30.0
5.	Thadoda	Thikhi	—do—	34.0
6.	Javada	—do—	—do—	139.0
7.	Bara-Nayagaon	—do—	—do—	130.00
8.	Naya Nagar	Nayanagar	—do—	50.0
9.	Khadipur	Damti	—do—	100.0
10.	Berissal	—do—	—do—	50.0
11.	Berisal	—do—	—do—	80.6
12.	Thadoda	Tikhi	—do—	10.8.7
13.	Thadoda	—do—	—do—	63.7
14.	Pachanpura and Dobia	Saimala	—do—	720.0
15.	Kishan Niwas	—do—	—do—	152.0
16.	Rampuria	Damit	—do—	50.0
17.	Kishnavato Ka Jhopara	Thumsa	—do—	25.0
18.	Govindpura	—do—		25.0
19.	Ganeshpura	Damjal		100.0
20.	Undroka Khera	Dhanwada		240.0
	Total			2382.5

Source : D.F.O; Bhilwara, 1997

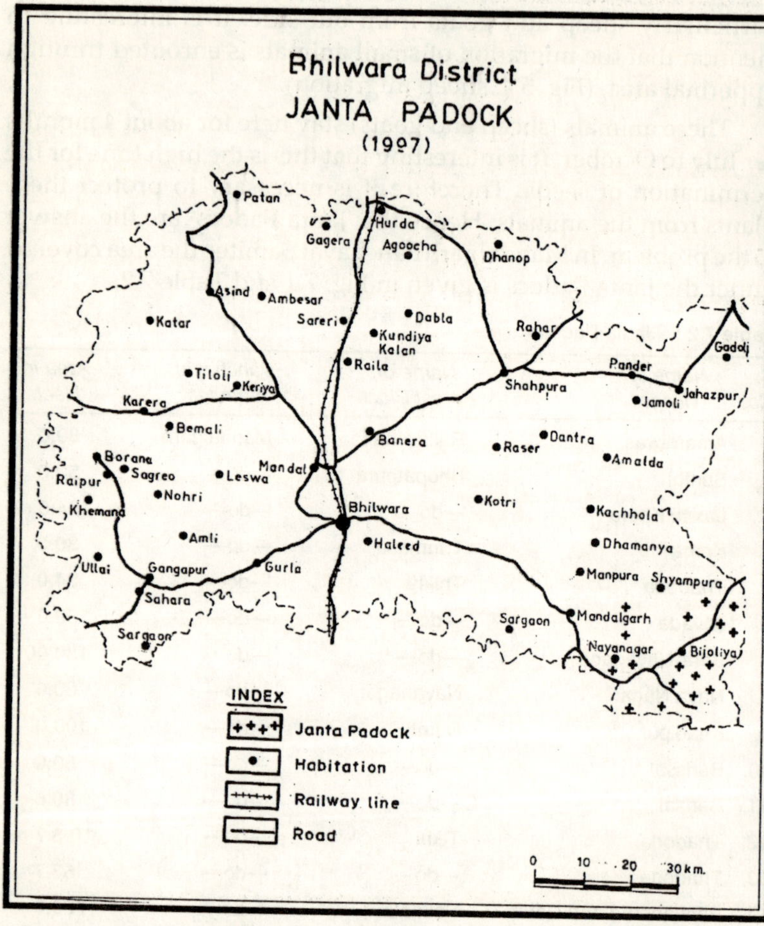

Fig. 7.1

Responses of the People and Social Forestry

In the beginning there was a wide spread resentment among the people for the development of social forestry. By 1981 there was a bitter agitation by the people against the authority . Some of the panchayats came forward but they had to face the consequences. But by late 1985 people realised the importance of social forestry and gradually came forward for the successful achievement of the scheme.

The real targets and the achievements of the community land plantation from 1983 onwards is shown in the Table 7.3.

Table 7.3 : Community Land Plantations from 1983 to 1993

S. No.	*Name of panchayat samiti*	*Total community land (in ha.)*	*Year wise plantation on community land in (ha.)*											
			1983	*84*	*85*	*86*	*87*	*88*	*89*	*90*	*91*	*92*	*93*	*Total*
1.	Asind	13529	—	80	—	50	60	30	85	175	—	50	—	530.0
2.	Banera	9585	35	—	16.7	20	52	24	65	55	—	45	—	312.7
3.	Hurda	6706	—	10	13.3	—	40	45	106	25	10	20	—	269.3
4.	Jahajpur	10566	—	—	—	—	—	60	—	30	—	—	—	90.0
5.	Kotri	11641	—	—	—	—	50	95	80	—	—	35	25	285.0
6.	Mandal	17469	—	—	—	—	—	30	18	65	167	70	34	384.0
7.	Mandalgarh	12426	—	—	—	—	—	455	174	—	—	—	—	629.0
8.	Raipur	6047	—	—	70	60	—	—	27	70	55	100	—	382.0
9.	Sahada	8365	—	60	—	50	43	65	90	70	140	40	50	758.7
10.	Sahapura	12824	150.7	60	80	140	60	25	277	—	93	20	96	851.0
11.	Suwana	10654	278.3	57	20	—	55	2.5	67	170	145	135	25	954.8
	Total	119812	464.0	267	200	320	360	831.5	989	660	610	515	230	5446.5

Source : District forest office, Bhilwara

It is evidently clear from the above table that in different years community land plantation, though launched on a very small area, but has a significant bearing. It is significant because the area formerly was lying wide open for without of any use. This land was considered as wasteland, but due to community land plantation scheme, may be on some part, was brought under some use to get fodder and fuel for the people. The highest land under community land was covered in Suwana by 954.8 ha., Sahapura by 851. ha. and Sahada by 758.7 ha. The lowest being in Jahazpur by 90 hectares.

The particular species of plants were favoured in different panchayat samitees like in Suwana panchayat samitee favoured by Babul, (Acacia nilotica) Vilayati Babul (Prosopis juliflora) and Israili Babul (Acacia tortlis) Sahapura panchayat samitee favoured by Israili Babul (Acacia tortlis) Babul (Ac. nilotica) and Vilayati Babul (Pr. Juliflora) in Sahada panchayat samitee favoured by Babul (Ac. nilotica), Vilayati Babul (Pr. juliflora) and Ronj (Ac. leucophloea) while in Raipur panchayat samitee is favoured by Vilayati Babul (Pr. juliflora), Babul (Ac. nilotica) and Ronj (Ac. leucophloea).

It is interesting to note that in most of the panchayat samities Babul, Vilayati Babul, Israili Babul and Ronj are the most common species except in dry areas where Pr. juliflora and Ac. senegal takes place. To better understand this fact that how people's responses have been in different parts, is indicated through some of the case studies of the villages listed in Table 7.4.

Unlike community land plantation the people's response may be analysed from the farm forestry also. Because in farm forestry people's participation is directly envisaged by digging pits, transporting, planting, and nurturing of plants. Under this scheme the achievements were sought as under (Table 7.5).

Impact Assessment

It is interesting to note here that the area covered by the forest department under community land plantation was estimated about 5446.5 ha., while the area covered under farm forestry is estimated to be just doubled *i.e.* 10465.5 ha. Thus 'Forest Department' did nothing except to provide the saplings. It reflects the people's responses and the interest in social forestry.

Table 7.4 : Survival Percentage and Growth Study

S. No.	*Name of plantation*	*Pancha-yat Samiti*	*Year of planting*	*Area in ha.*	*Saplings planted (in 000)*	*Scheme*	*Year of evaluation*	*Survival %age*	*Av. Height in mts.*	*Av. girth in mits.*
1.	Kherbad	Suwana	1981	32	12.8	F. W. P.	1995	78	5.0	0.45
2.	Kherbad	Suwana	1987	30	50.4	F. W. P.	1995	82	4.5	0.30
3.	Kherbad	Suwana	1989	20	4.0	S. P. P.	1995	77	4.0	0.30
4.	Kherbad	Suwana	1990	35	38.5	F. W. P.	1995	79	3.5	0.25
5.	Kherbad	Suwana	1992	50	40.0	F. W. P.	1995	81	3.0	0.20
6.	Mangrop	Suwana	1990	50	66.0	F. W. P.	1995	78	4.0	0.36
7.	Mangrop	Suwana	1991	15	16.5	T. P.	1995	76	4.5	0.35
8.	Gadarmala	Suwana	1983	43	17.2	F. W. P.	1995	72	5.0	0.40
9.	Gadarmala	Suwana	1989	20	2.0	S. P.P.	1995	76	4.0	0.35
10.	Jabarkia	Asind	1990	30	6.0	S. P. P.	1995	74	3.0	0.20
11.	Jabarkia	Asind	1990	30	33.0	F.W.P.	1995	71	3.0	0.20
12.	Bankakhera	Kotri	1988	50	55.0	F. W. P.	1995	56	2.5	0.20
13.	Bankakhera	Kotri	1988	15	1.5	S. P. P.	1995	45	2.0	0.20
14.	Khamor	Shahpura	1989	35	56.0	F. W. P.	1995	72	4.0	0.25
15.	Khamor	Shahpura	1991	50	35.0	F. W. P.	1995	69	3.0	0.20
16.	Khamor	Shahpura	1992	50	40.0	F. W. P.	1995	78	2.0	0.15
17.	Potlan	Sahada	1986	50	20.0	F. W. P.	1995	67	4.5	0.35
18.	Potlan	Sahada	1991	10	7.0	F. W. P.	1995	81	3.5	0.20
19.	Potlan	Sahada	1991	30	33.0	F. W. P.	1995	75	2.5	0.20
20.	Potlan	Sahada	1993	10	2.0	F. W. P.	1995	80	2.0	0.15
21.	Potlan	Sahada	1993	10	11.0	F. W. P.	1995	72	2.5	0.15

Source : D.F.O., Bhilwara and based on field survey of plantations.

Table 7. 5 : Planting under Farm Forestry Scheme

S. No.	Year	Physical Target in Lacs	Achievement in lacs	Equivalent planting area in ha.
1.	1983-84	8.00	11.97	598.5
2.	1984-85	18.00	15.27	763.5
3.	1985-86	17.00	18.54	927.0
4.	1986-87	22.00	22.22	1111.0
5.	1987-88	20.00	26.31	1315.5
6.	1988-89	35.00	27.94	1397.0
7.	1989-90	33.25	24.20	1210.0
8.	1990-91	27.50	19.18	959.0
9.	1991-92	27.00	27.03	1351.5
10.	1992-93	15.50	16.65	832.5
	Total	223.25	209.31	10465.5

Source : District forest office, Bhilwara.

Economic aspects may be considered important in the impact assessment. In fact it is the money makes the way. If we look to the yearwise budget allocation and the money spent on the social forestry schemes alongwith some of the forestry allied acitivities. We find that in different years. It has a different projections. In total from 1986 to 1993 about 806.64 lacs rupees were allocated for the development of social forestry in the district. In different parts in the same period the money spent, is equally good enough of the allocation, which was covered for the 92.38 lac plants in the district, (Table 7.6).

Table 7.6 : Yearwise Budget Allocation and Money Spent Under Various Forestry Schemes

S. No.	Year	Area under plantation in ha.	No. of plants in lacs	Budget allocation in lacs	Money spent in lacs.
1.	1986-87	1495.0	6.58	58.32	58.32
2.	1987-88	1070.0	11.98	115.69	115.69
3.	1988-89	1626.5	11.26	71.27	71.27
4.	1989-90	1805.0	14.29	119.57	119.57
5.	1990-91	2160.0	16.88	110.75	110.75
6.	1991-92	1935.0	11.67	162.15	162.15
7.	1992-93	1550.0	13.51	113.36	113.36
8.	1993-94	815.0	6.21	55.53	55.53
	Total	12456.5	92.38	806.64	806.64

If we go through the individual schemes carried on along with the social forestry are like R.D.F. (State Plan), compensatory afforestation, etc. For these projects budget was allocated about 55.53 lac rupees in 1993-94 (Table. 7.7).

Table 7.7 : Money Spent on Different Schemes and Area Under Plantation.

S. No.	*Name of Scheme*	*1992-93*			*1993-94*		
		Area in ha.	*Budget*	*Money spent*	*Area in ha.*	*Budget*	*Money spent*
1.	N.S.F.P.	530	72.75	72.75	490	46.68	46.68
2.	Fuelwood	800	31.13	31.13	50	1.50	1.50
3.	R.D.F. State Plan	100	3.88	3.88	—	—	—
4.	Compensatory aforestation	120	5.60	5.60	—	—	—
5.	Soil conser-vation	—	—	—	250	5.17	5.17
6.	Energy plantation	—	—	—	25	2.18	2.18
	Total	155	113.3	—	815	55.53	55.53

Source : District forest office, Bhilwara.

Further it is interesting to have a look to the estimated income generation of forest products in money value. The case studies made for the purpose indicates a bit amount of income in 4 panchayat samities, particularly in some villages falling under them (Table 7.8).

Table 7.8 : Estimated Annual Income Generation of Forest Products.

S. No.	*Name of planta-tion*	*Year*	*Panch-ayat Samiti*	*Area in ha.*	*Grass quan-tity in tonnes*	*Pro-duce cost in Rs.*	*Fuel-wood quan-tity in tonnes*	*Pro-duce cost in Rs.*
1.	Kaliyas	1990	Asind	25	5.0	5000	2.0	2000
2.	Jabarika	1990	Asind	30	6.0	6000	7.0	7000
3.	Raila	1990	Banera	20	8.0	8000	2.0	2000
4.	Beran	1990		30	20.0	20000		
5.	Kanvaliyas	1989	Hurda	15	9.0	9000	2.5	2500
6.	Jasoria	1992	Hurda	20	7.5	7500		
7.	Thabola	1992	Mandai	35	28.0	28000	10.0	10000
8.	Luharia	1992	Mandal	35	26.0	26000	3.5	3500

Source : On the basis of field study.

It is estimated that Kaliyas and Jabarkia village of Asind panchayat samitee produces about 11 tons of grass and 9.0 tons of fuel wood generating income about Rs. 11000 and Rs. 9000 respectively. Similarly Luharia and Thabola village of Mandal Panchayat Samiti producing 54 tons of grass and 13.5 tons of fuel wood annually generating income of about Rs. 54000 and Rs. 13500 respectively but has a greater bearing for the villagers in particular.

It is interesting that this income has been generated out of the waste land lying vacant for years without of any use. In that way it may be said that if all the waste lands lying vacant in different village sites, if brought under such pretty uses it may generate a good amount of income for the villagers.

Thus the economic aspects play an important role in changing the economic structure of the village people, and hence the very purpose of the social forestry too, is served. Thus it has a greater impact.

From this fact it may be realised that the people's participation, their interest and enthusiasm has broken all the barriers and have come forward for the greater benefits of the forest resources.

It is noted that they need some facilities and guidance. It is to see that they are treated with a better guidance and expertise for the better achievements. Therefore the real impact assessment is on the way when this scheme is exercised on a larger scale. Because at present the area covered under social forestry is limited and restricted from different point of views.

Cultural Constraints

Any scheme has to pass through a critical social constraints. In traditional societies like ours, it is very difficult to lauch a scheme without opposition at the beginning, but as soon as the fruits are available it too, becomes difficult to check it on.

Some times social groups and their religion, caste, tradition, customs, taboos, etc., play vital role in putting checks on many useful practices. Villagers do not readily accept these changes unless and untill realise the benefits.

Similar case has been happened with the farm forestry and social forestry. When these schemes were started, villagers thought that the government is grabing or seizing their land. But later on they realised the importance of social forestry and farm forestry and favourably responded to the schemes.

References

1. Arora Hema and Khara, Arvind, 1993, People's participation Society for promotion of waste lands development (S.P.W.D.). Vol. VIII No. 4.
2. Project Paper. 1985, India National Social Forestry, Project (386-0495) Vol. II.
3. Kapoor, R.P. Chose Abhijit and Upadhyaya, A.K. 1997, Status of Joint Forest Management in Rajasthan. Executive Development Programme, P. 2. Forestry Training Institute, Jaipur.

Chapter-8

Future Strategy

Awareness About the Environment

Amidist the challenges and crises that we are facing, today participatory management of natural resources offers a hope to the present problem. Future strategy in that sense makes an important point of participation of people to generate resources from the forests. This can be done with the effective and scientific management approach.

Social forestry in this respect enjoys a far better resource management basis for the people's participation point of view. It is to provide opportunities to men and women and marginalisation prone groups, local culture conservation, preparation of management plans, linking of income generating programme with conservation, livelihood needs of the people and role of non governmental organisations (N.G.O.s) in participatroy development.[1]

Participatory forestry has several features. Grass and fodder base economy needs different kind of strategies to be followed for forest regeneration.

In a nutcell we may state that all these programmes are accorded in order to facilitate environmental conditions to remain smooth. Thus environment is the key to the whole. Because if the forest cover is removed or degraded, it has a counter effect on man, animal and other living beings. Therefore the idea behind the social forestry is to involve people to maintain their eco-system and to meet their needs simultaneously. The idea of coexistence is basically the same that people's participation in order to make awareness about the environment in which they are living.[2]

National forest policy (N.F.P.) 1988 envisages that 33 per cent of the geographical area should be under forests or treecover. National forest policy also envisages over riding consideration for environmental concerns against commercial priorities. The state forest policy has not been officially announced but for all practical purposes various strategies, plans and programmes are taken up within the overall framework of the N. F. P. with special emphasis on combating desertification, checking soil erosion, and denudation with massive people's movement.

The present forest area of the state is 9.32 per cent and of the district is only 7.60 per cent of their respective geographical area. Which is far below the national average of 23 and 33 per cent as envisaged by National Forest Policy 1988. Considering the existing extent of forest area and availability of non forest waste lands, about 20 per cent of the district's geographical area can be brought under forests or tree cover by adopting appropriate technologies and providing required level of investments. The existing area under various protected area network (PAN) is 2.8 per cent of the state's geographical area, which compares favourably with that recommended by the wild-life institute of India, for Rajasthan which envisages 3.1 per cent of the geographical area of the state under PAN. The state has proposed world heritage sites like Sambhar lake, other wetlands, sacred groves locally called orans or Gochars and other similar area rich in biodiversity, is a monument of our rich cultural hertiage. On a conservative estimates, these area constitute about 1.8 per cent of the geographical area. Considering all these aspects, it is possible to have 5 per cent of geographical area of the state for biodiversity purpose.[3]

Thus, the future strategy needs a developmental approach. In order to effectively tackle the problem of land of different nature, site specific strategies are to be adopted. In view of the needs and aspirations of the state and country in general and the local people in particular. For this purpose participatory management approach and Watershed/Cluster approach are considered important.

Forest Ranges in Bhilwara District

As per the government forest policy and for management point of view the total forest area of Bhilwara district is divided in to 6 "Forest Ranges". For the purpose of administrative plan perspective, it is a better, classification. For this purpose all the 6 units are divided into R.F., P.F. and U.F. (Unclassed forests) with their number of forests blocks (Fig. 8.1 and Table 8.1)

Table 8.1 : Range wise Forest Area in Bhilwara District.

(Area in ha.)

S No.	*Name of range*	*R. F.*	*P. F.*	*U. F.*	*Total block/area*
1.	Asind	9/2156	27/1983	45/5024	81/9163
2.	Bhilwara	22/2264	16/1151	32/3081	70/6496
3.	Gulabpura	4/851	2/71	—	6/922
4.	Jahazpur	29/8207	51/8117	1/56	81/16380
5.	Mandalgarh	37/29940	28/13205	3/155	68/43300
6.	Shahpura	—	31/3030	2/127	33/3157
	Total	(101)/43418	(155)/27557	(83)/8443	(339)/79418

Source : D.F.O. Bhilwara, 1996.

Note : Figures in parenthesis denotes the number of "Forest-Blocks".

The table indicates that the total area covered under "Forest Ranges", is about 79418 hectares, with 339 forest blocks. These blocks with their different area under different categories are better shown in the Table 8.1. This reflects that Gulabpura forest range occupying lowest area 922 hectares which is about 1.1 per cent of the total forest of the district. This case may be regarded as the administrative anamoloy in forest classification. Whereas Mandalgarh Range occupying near about 55 per cent area of the total forests of the district. It needs to readjustment of the forest ranges.

Economic Constraints and Hardships

Looking to the demand and supply of fuel, fodder and other items from the forests is a bit complicated matter. It projects a different projection of need and supply of different items. If we look to the availability of fuel and fodder and their needs in the Table 8.2 alongwith the achievement of plantations as a future strategy, we find a different projection. The total fuel available in the district is reported as about 357.3 thousand tonnes. This shows

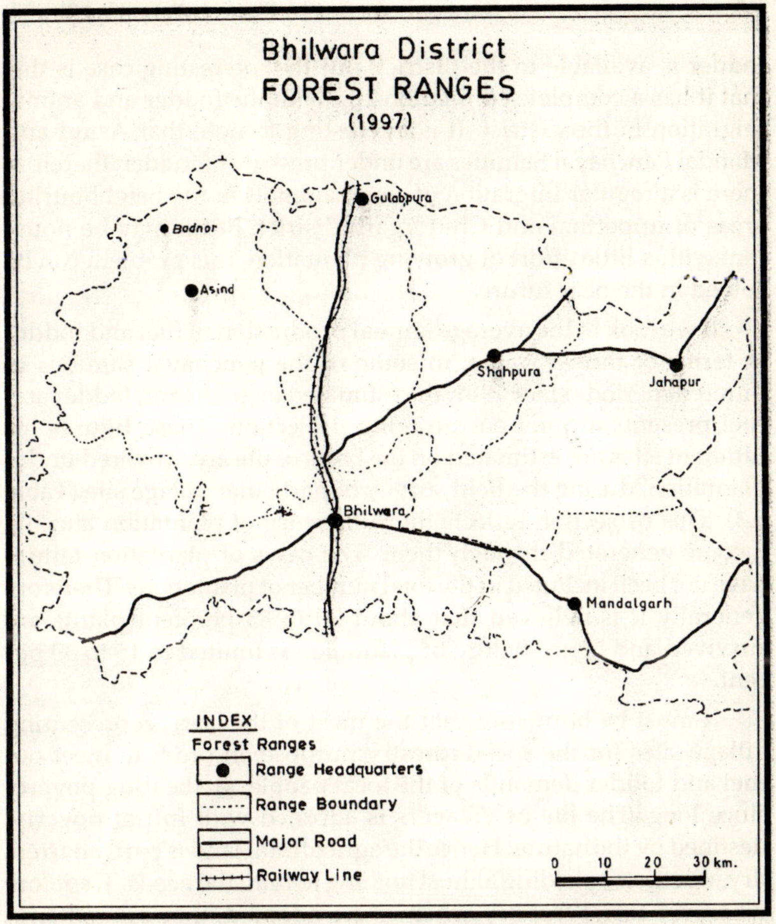

Fig. : 8.1

that 62.5 per cent fuel is available in the district of the total need. That means about 40 per cent fuel is needed to fulfil the requirements. It may also be noted that the fuel made available from different sources like L.P.G. and kerosene is excluded from the present requirement. The local projection in different panchayat samities is a bit different, as Suwana panchayat samiti needs 104.5 thousand tonnes of fuelwod simply because of the industrial use, particularly of "Process Houses".

Similar is the case with the fodder availability and its need. At present nearly 78.4 per cent or 2617.5 thousand tonnes of the total fodder is available in the district. But the interesting case is this that it has a complete co-relationship with the fodder and animal migration in the district. It is interesting to note that Asind and Mandal Panchayat Samities are under pressure of fodder, therefore there is a regular migration of small animals in the neighbouring areas of uppermal and Chittorgarh district. But it may be noted that with a little effort of growing plantation, this problem can be solved in the near future.

If we look to the average annual production of fuel and fodder in terms of money value, in some of the panchayat samities in different periods since 1981, the estimated income from fodder and fuel presents a quiet encouraging projection. These figures for different sites are estimated on the basis of the area covered under plantation, during the field survey of particular village site (Table 8.3). This projection reflects the sufficiency of plantation and the income generated through them. The cases of plantation failure have not been included in the total number of plantations. Therefore generally it is believed that about 70 to 85 per cent plants are survived and hence failure of plantation is limited to 15 to 30 per cent.

It must be borne out that the most of the areas representing village sites for the social forestry purpose, in order to meet out fuel and fodder demands of the local people, are bearing poverty since long. The life of villagers is adorned with full of poverty, destined by the nature. Hence the agricultural land is partly barren, dry, waste etc., yielding almost nothing to fulfil the needs. Therefore wherever social forestry schemes are adopted, they are either in poor condition or with uncultivated waste lands. Hence, economic hardships are posed frequently.

Employment Generation

From the future strategy point of view the different schemes under the social forestry, indirectly speaks of the employment generation also. If we look to the plantation schems of agroforestry, farm forestry etc., they indirectly generate the employment for the village people. First of all the plantation schemes needs human labour for the fencing, digging of pits, plant raising and planting, watering to the plants, weeding and hoeing and finally tendering and looking after the plants etc. Generally it is observed by the

Table 8.2 : Future Projection or Strategy of Social Forestry

S.No.	Panchayat Samiti	Available		Annual need		Achievements in plantations		Future targets
		Fuel (in 000 tonnes)	Fodder (in 000 tonnes)	Fuel (in 000 tonnes)	Fodder (in 000 tonnes)	No. of plants (in lacs)	Area in ha.	Area to be brought under plantation (in 000 ha.)
1.	Asind	42.6	253.4	60.9	362.0	4.765	530.0	9.15
2.	Banera	17.7	88.6	32.2	126.6	3.184	312.7	7.25
3.	Hurda	16.9	112.0	33.7	160.6	2.319	269.3	8.40
4.	Jahazpur	37.8	258.0	54.0	344.0	0.990	90.0	8.10
5.	Kotri	24.5	249.9	44.5	312.4	3.000	285.0	10.00
6.	Mandal	32.9	225.4	59.8	346.8	3.753	384.0	13.45
7.	Mandalgarh	80.0	500.0	66.6	457.7	1.504	629.0	3.65
8.	Raipur	14.4	129.5	26.1	215.8	3.795	382.0	5.85
9.	Sahada	18.3	190.0	36.6	326.7	7.097	758.7	9.15
10.	Shahpura	26.4	340.0	52.7	325.9	8.731	851.0	13.15
11.	Suwana	45.8	270.7	104.5	360.9	9.790	954.8	29.35
	Total	357.3	2617.5	571.6	3339.4	48.928	5446.5	117.50
Percentage availability		62.5%	78.4%					

Source : On the basis of field survery.

Table 8.3 : Projection of Fuel and Fodder production.

S. No.	Name of plantation	Pancha-yat Samiti	Year of planting	Area in ha.	Saplings planted (in 000)	Scheme	Year of evaluation	Estimated average annual production: Fuelwood Produce (in tonnes)	Fuelwood Value (in ooo Rs.)	Fodder Produce (in tonnes)	Fodder Value (in 000 Rs.)
1.	Kherabad	Suwana	1981	32	12.8	F. W. P.	1995	15	15	20	20
2.	Kherabad	Suwana	1987	30	50.4	F. W. P.	1995	6	6	15	15
3.	Kherabad	Suwana	1989	20	4.0	S. P. P.	1995	3	3	20	20
4.	Kherabad	Suwana	1990	35	38.5	F. W. P.	1995	5	5	15	15
5.	Kherabad	Suwana	1992	50	40.0	F. W. P.	1995	—	—	20	20
6.	Mangrop	Suwana	1990	50	66.0	F. W. P.	1995	6	6	50	50
7.	Mangrop	Suwana	1991	15	16.5	T. P.	1995	2	2	15	15
8.	Gadarmala	Suwana	1983	43	17.2	F. W. P.	1995	10	10	20	20
9.	Gadarmala	Suwana	1989	20	2.0	S. P.P.	1995	2	2	25	25
10.	Jabarkia	Asind	1990	30	6.0	S. P. P.	1995	2	2	35	35
11.	Jabarkia	Asind	1990	30	33.0	F.W.P.	1995	7	7	10	10
12.	Bankakhera	Kotri	1988	50	55.0	F. W. P.	1995	5	5	25	25
13.	Bankakhera	Kotri	1988	15	1.5	S. P. P.	1995	—	—	20	20
14.	Khamor	Shahpura	1989	35	56.0	F. W. P.	1995	5	5	17	17
15.	Khamor	Shahpura	1991	50	35.0	F. W. P.	1995	6	6	30	30
16.	Khamor	Shahpura	1992	50	40.0	F. W. P.	1995	—	—	20	20
17.	Potlan	Sahada	1986	50	20.0	F. W. P.	1995	6	6	15	15
18.	Potlan	Sahada	1991	10	7.0	F. W. P.	1995	1	1	15	15
19.	Potlan	Sahada	1991	30	33.0	F. W. P.	1995	3	3	20	20
20.	Potlan	Sahada	1993	10	2.0	F. W. P.	1995	—	—	6	6
21.	Potlan	Sahada	1993	10	11.0	F. W. P.	1995	—	—	5	5

Source : Based on village field survey. Figures are estimated on the basis of production.

forest department that one hectare of community land plantation provides about 400 man days.

Similarly, if fodder in general and grasses in particular are grown in an area, the live-stock based enterprise system automatically generates the multipurpose economy for the people. Some people may start live-stock breeding or may take interest in dairy or meat and hence are engaged in allied activities. Similarly if the forest products are raised in variety, people may start a sort of business on produced items and hence income is generated for them.

Thus, social forestry aims to raise or to generate income for the village poorers alongwith the major plantation schemes. Thus it fulfils the wide gap between urban and rural masses. Actually it is a way to the further development of cottage industry. It may generate income from different sources like toy industry, basket and broom making etc.

Wood Based Industry

This industry since long is perishing on timber wood, directly coming from the forest areas in the vicinity. Therefore it would not be wrong to say that in the exploitation of forest areas, this industry has played a major role. It is in otherwise forests provides the basic need of the timber. Generally if we want to get timber from the forests we have to cut the tree, while it is matured. Thus one has to take the note that timber wood industry exploits the forests. In order to maintain the balance between forests and supply of wood, it is necessary to seed a plant first in lieu of felling out of the tree.

For the present situation the timber wood industry is passing through a critical phase of its development, as government is imposing serious checks and restrictions over the cutting down of trees. The precious wood like sandalwood, rose wood and teak etc. are completely banned. It is illegal to cut any green tree, therefore from the ecological point of view, it is a wiseable step, to maintain perfect eco-system.

Future Strategy

Whatever plantation has been done in the past is the true picture for the future. It may be taken that the social forestry scheme was started in Bhilwara district in 1980, since then, there has been different kind of plantations in different parts of the district. In the beginning there was a direct target of plantation in areas where panchayat land was available. In doing so it was not a consideration that what plant species is suitable to an area. It is generally believed that the area as a whole belongs to a particular type of species like Prosopis juliflora, Ac. Nilotica etc. Hence no other species is suitable.

Under this assumption a wide range of plantation work at different village sites was started and a particular type of tree species, were planted. It was under the impression that a particular tree species is easy to cultivate and needs less maintenance as compared to any other hardy plant species.

This practice brought some serious repercussions among villagers; there were some serious cases of particular plant species which actually did not suit to the area *e.g.* P. Juliflora (Vilayati Babool) and Eucalyptus (Safeda). The most suitable plant species are Babul (Ac. nilotica), Ronj (Ac. leucophloea), Ber (Zizyphus jujuba) etc. These speices are important from the economic point of view also. The former case is suitable only in areas of community waste lands.

Keeping this view in mind for the current year 1997 a plan perspective has been drawn for the plantation of tree species in suitable village panchayats (Table 8.4).

In this table it may be noted that the most of the tree species are indigenous. The plantation of P. juliflora is reduced to 16 per cent but the area under Desi Babul is raised upto 46 per cent. The area under some of the particular tree species which are widely in demand was negligible till 1990. But looking to the importance of these tree species (Bamboo Anwala, Khair, etc.) the area has been widen.

At the beginning Eucalyptus was cultivated on wide range of areas under the impression that it is money fetching tree. But it is only after the losses of land and revenue, people realized the environmental unsuitability of the plant. It needs more water hence lowered the ground water table in areas of plantation. It provides neither fodder nor fuel and nor even shade. Now this plant is replaced by Neem, Shisham, Siras and Karanj etc.

Table 8.4. Species wise Plant Raising Information for 1997 Planting, in Bhilwara District

Sl. No.	*Name of planting site*	*Scheme*	*Area in ha.*	*Babul*	*Kumata*	*Ber*	*Ronj*	*Vil. Babul*	*Khari*	*Avenue plants*	*Bamboo*	*Anwla*	*Siras*	*Shisham*	*Others*	*Total*	*Actual planting targets*
1.	Asuna	S. P.	20	2900	750	500	—	—	—	250	—	—	—	—	—	4400	4000
2.	Khemana	F. W. P.	30	28800	2500	—	2500	2500	—	—	—	—	—	—	—	36300	33000
3.	Galiyawadi	S.P.	30	5600	—	800	100	—	—	100	—	—	—	—	—	6600	6000
4.	Sahada	F. W. P.	30	29800	—	500	—	5500	—	500	—	—	—	—	—	36300	33000
Total Bhilwara			110	67100	3250	1800	2600	—	—	850	—	—	—	—	—	83600	76000
5.	Basotha Ki Dhar	RDF-I	50	11000	—	3300	5500	—	7700	—	5500	5500	—	—	—	38500	35000
6.	Bhopatpura	RBH	50	4400	—	8800	5500	—	29700	—	—	11000	1100	—	—	60500	55000
7.	Damti	RDF-I	50	5500	—	11000	11000	—	5500	—	—	5500	—	—	—	38500	35000
8.	Genoli	RBH	50	11000	11000	5500	5500	—	16500	—	5500	5500	—	—	—	60500	55000
9.	Nahargarh	RDF-I	50	7700	4400	6600	8800	—	11000	—	—	—	—	—	—	38500	35000
10.	Gowatamata	RDF-II	50	4000	—	—	—	—	7000	—	—	—	—	—	—	11000	10000
Total Mandalagarh			300	43600	15400	35200	36300	—	77400	—	11000	27500	1100	—	—	247500	225000
11.	Ganeshpura	CSS	40	30800	—	—	2200	—	—	—	—	2200	—	—	—	35200	32000
12.	Thabola	CSS	40	24200	—	—	1650	—	—	—	—	550	—	—	—	26400	24000
13.	Banera-I	RDF-I	50	22000	5500	—	4950	5500	—	—	—	550	—	—	—	38500	35000
14.	Banera-II	RDF-II	50	2200	3850	—	1100	3300	—	—	—	550	—	—	—	11000	10000
15.	Karanti	RDF-II	45	2200	4400	—	—	3300	—	—	—	—	—	—	—	9900	9000
16.	Sanodiya	F. W. P.	30	33000	—	—	—	1100	—	—	—	2200	—	—	—	36300	33000
17.	Dhanpura	S. P.	50	9900	—	—	—	—	—	—	—	1100	—	—	—	11000	10000
Total Gulabpura			295	124300	13750	—	9900	13200	—	—	—	7150	—	—	—	168300	153000

Sl. No.	Name of planting site	Scheme	Area in ha.	Babul	Kum-ata	Ber	Ronj	Vil. Babul	Khari	Avenue plants	Bam-boo	Anwla	Siras	Shi-sham	Others	Total	Actual planting targets
18.	Fathepura	CSS	40	19800	8250	1100	5500	—	—	—	—	—	—	—	—	35200	32000
19.	Ropan	CSS	25	18800	—	500	2500	—	—	—	—	200	—	—	-	22000	20000
20.	Badla (Kajoria)	CSS	35	25800	—	1100	3300	—	—	600	—	—	—	—	—	30800	28000
21.	Bland (Amli)	F. W.P.	30	23100	—	550	2530	—	—	—	—	220	—	—	—	26400	24000
22.	Amli (Sangriya)	F. W. P.	17	17000	—	—	2500	—	—	—	—	200	—	—	870	20570	18700
Total Shahpura			147	104500	8250	3250	16330	—	—	600	—	620	—	550	870	134970	122700
23.	Rajpura	RDF-I	65	37150	—	1500	10000	—	—	400	1000	—	—	—	—	50050	45500
24.	Jerdamata	RDF-I	35	3800	15400	—	7750	—	—	—	—	—	—	—	—	26950	24500
25.	Kaliyas	FWP	23	22000	—	2750	2750	—	—	330	—	—	—	—	—	27830	25300
26.	Badnagar	RDF-II	55	—	12000	—	—	—	—	—	—	—	—	—	—	12000	11000
Total Asind			178	62950	27400	4250	20500			730	1000				—	116830	106300
27.	Kheruna (Titoda)	RBH	50	32000	—	11000	8000	8000	—	—	1500	—	—	—	—	60500	55000
28.	Gauramgarh	RBH	50	—	5000	1000	2000	52000	—	—	500	—	—	—	—	60500	55000
29.	Dhandhola	RBH	50	—	17000	1000	9000	27500	—	—	500	—	—	—	—	55000	50000
30.	Chandmari	RBH	50	—	20700	2700	6000	30000	—	—	1100	—	—	—	—	60500	55000
31.	Itunda	RDF-I	50	—	16000	2000	2500	20000	3000	—	500	—	—	—	—	44000	40000
32.	Kheruna	RDF-I	50	15000	—	5000	12000	3500	—	—	2000	1000	—	—	—	38500	35000
33.	Dhamaniya	FWP	20	10000	—	5000	—	—	—	—	5000	4000	—	—	200	24200	22000
34.	Adamaya Pariya	RDF-II	50	4000	—	—	—	7000	—	—	—	—	—	—	—	11000	10000
Total Jahazpur			370	61000	58700	27700	39500	148000	3000	—	11100	5000	—	—	200	354200	322000
Grand Total			1400	463450	126750	72200	125130	169200	80400	2180	23100	40270	1100	550	1070	1105400	1005000

Therefore social forestry is not merely a plantation scheme but it needs utmost care and judicious selection of tree species and their development in particular area. This point stress the need of future strategy.

For the future projection it may be taken that if we want to launch a future strategy for the social forestry schemes in Bhilwara district, the following category of lands may be made available (Table 8.5).

Table 8.5 : Different Category of Lands Available for. S.F.

S. No.	*Category of land*	*Area available*
1.	Forest land	35000 hectare
2.	Revenue waste land	40000 hectare
3.	Panchayat land	20000 hectare
4.	Private land	22500 hectare
	Total	117500 hectare

The development of the above land categories needs about 150 crores, of rupees, for the fifteen years of plan. It may generate extra benefit of employment in rural areas, alongwith grasses, legumes fodder, fuelwood and thatching materials etc.

From the above discussions and some of the analytical explantations, it brings out that social forestry has a wider scope for the future strategy in many respects. It maintains the environment, eco-system on one side and generate employment, income and fulfilment of human needs on the other. In this respect the social forestry needs wider care and protection from both the sides; government and the people.

References

1. Govil, D. P. and Pandey, D. N. 1997, Participatory forestry in Aravallis. Selected readings. Executive Development Programme P. 18. Forestry Training Institute Jaipur.
2. Kapoor, R. P., Ghose, Abhijit and Upadhyaya, A. K. 1996, Status of Joint Forest Management in Rajasthan. Executive Development Programme, P. 2 Forestry Training Institute Jaipur.
3. Forest Department, Government of Rajasthan, Jaipur, 1996, State Forestry Action Programme, (1996-2016), P. 30.

Chapter-9

Problem and Prospects

While concluding the whole script, it would be relevant to discuss some of the problems and prospects of the forestry systems in perspective background of Bhilwara district. As far the problems are concerned, it is very appropriately discussed in the introduction part that there has been growing gap between demand and supply of forest resources in the rural areas. Similarly there is an increasing degradation of forests wealth, due to unwarranted exploitation of resources, which led to the environmental problems and hence the ecological problems. The degradation of environment and ecological balance is more serious problem, for which no remedy was forseen.

It is no wonder that the state remained with hardly 9.32 per cent of forests whereas the district is reported about 7.6 per cent. The actual forest cover is much more lesser. While as per the international forums and the N.F.P. (National Forest Policy) demands at least 33 per cent of the total area should be under forests.

Looking to the above analogy the problem is more serious where a meagre land is matched with the forest cover.

To meet out the growing challenges of the environment and man's relationship and the environmantally thwarted programmes by the government at both the levels *i.e.* state and village. Some of the programmes implemented by the state government are as follows :

1. Reclamation of Problem Lands
2. Agroforestry System

3. Management of Forest, Plantation through J. F. M./ P. F. M.
4. Grazing Management
5. Bio-diversity Management
6. Forestry Extension
7. Combating Desertification
8. Watershed/Cluster Approach of Afforestation
9. Forestry Research etc..... etc.

But the major problem in rural areas, still looming at large, was of the fodder and fuel. In fact down cutting of forest trees was mainly due to the demand of fuel and fodder. Therefore it was realised, though at a latter stage, that the people's need of fuel and fodder must be fulfilled by the alternate sources, so that the further degradation of forests may be stopped.

With this idea in mind the S.F. Scheme was taken up by the national/State/ district/panchayat level authorities. It is nothing but a sort of people's participation in forestry development, for which much has been written in the preceding pages.

It is for the present relevance that the S.F. includes the Agroforestation, Rehabilitation of degraded forests (R.D.F.), (Rehabilitation of Barren Hills (R.B.H.), Village Fuel Wood lots (V.F.W.), Silvipastoral plantations (S.P.) Road and Canal side plantations etc. The Social Forestry is referred to the fulfilment of fuel and fodder, is better known as 'village wood lot'.

The systematic programme of community woodlot (Village woodlot) was started in Bhilwara around 1980, following introduction of the world bank who assisted the social forestry project in the state. Implementation experience of such programme has shown that panchayat response in creating 'Self-help' woodlot by providing inputs had been negligible due to financial constraints and the developed woodlots with government inputs turned out to be pure departmental activity in the absence of any effective involvement of the community or panchayats in the planning/ designing/ establishment process. Actually, the findings are that the villagers or panchayats are still reluctant in most of the cases to take up management responsibilities after the stipulated time period. In order to develop the commitment and managerial capacity of Management Committees (V.F.P. and M.C.), there is a need for interaction from the very beginning of such community woodlots programme, so that it becomes a self managed, self sustaining system of fuel and fodder production over harvesting cycles at the village or community level.

Though a management plan that reflects aspirations of the village community and responsibilities relating to such community land plantations (woodlots), was envisioned by the forest department. But it could not serve the purpose as an institutional tool for the proper management of such woodlots. Attempts should be made to formulate a simple and workable document that not only to focus towards true collaborative development through intensive interaction with the villagers, but also consider the level of the field staff's knowledge/skill and time available within the mandays of increasing physical target attainment responsibilities.

Intensive deliberations should be held with the villagers and panchayat members to ascertain community perception regarding purpose of raising such woodlots. The species selection and proposed management system as well as product disposal plan for attainment of objectives should be made clear to the V.F.P. and M.C., so that in future there remains no controversy between the department and the community. This will serve as a tool for achieving village co-operation and developing institutional capability overtime to come. This will ensure the active participation of the people in the programme and giving them specific idea about the product outturn and financial return based on their perception.

It would be relevant here to discuss the 'Janta Van Yojana' in the context of future prospects.

The Janta Van Yojana in fact is a shared responsibility between panchayats (village people) and the government. In which panchayat provides the land for the plantation sites. The government authorities with the help of district forest officer (D.F.O.) selects the species according to the site and requirement of the people, provides the other infrastructural facilities to the villagers. Thus it is villager's responsibility to look after and to protect the planted saplings.

Thus it may be said that Janta Van Yojana is a Joint Venture of recasting forests for the future and the fulfilment of the fuel and fodder.

Some of the Findings

While concluding the book, one may come to the following findings :

1. At the beginning there was no proper response to the social forestry by village panchayats, as villagers were unaware about the results of the social forestry.
2. Sites selected for some species like Babul, Israili Babul, Juliflora etc. were not proper for their growth and development. It was realized a little latter that a particular species should have been planted some where else.
3. Some of the sites selected for the Silvipastoral planatations were infact suitable for the fuelwood plantations and vice-*versa e.g.* Satiliyas village in Sahada Panchayat Samiti.
4. Migration of animals in different parts of the district specially in Madalgarh needs to be checked.
5. Fodder for the migratory herds should be developed at their local places.
6. Janta padocks should be developed in different parts, so that unwarranted grazing may be checked.
7. Social forestry's main emphasis is on the fuel and fodder and hence, cottage industries are little bit neglected. Therefore it is suggested that the cottage industries should be highlighted, wherever their related items are available.
8. It has been witnessed in many places in the district that whereas the area has been increased under the plantations, the regeneration of local species as well as wild animals also has been developed.
9. Study conducted in the district while assessing impact of community land plantation has shown that women prefer indigenous trees which are a source of fuel and fodder while men preferred trees that have market value and can earn some money.

Short Comings

Despite many good reasons of the social forestry in the state in general and district in particular, there are some short comings too.

1. There is a gap between people and government body to patch up the doubts.
2. Villagers are still susceptible about the government schemes.
3. It seems that authorities are not very much serious about the schemes, hence attitudinal change in the style of functioning of F. Dept. is needed. Due to this the department could not win the trust and confidence of the people.
4. People's participation was not secured in planning and regeneration efforts to ensure sustainability and equitable distribution of forest products from the regenerated (Community) lands, hence they are reluctant to take up the management responsibilities after the stipulated period.
5. Raising of seedlings and distribution is the starting point for most of the S.F. activities but in the district only quantity was given importance not the quality of the seedlings.
6. Seedlings once distributed or sold to the farmers and communities, Govt. departments and industries and Forest Dept. share no responsibility. Hence, there is no way to know, what has happened to the distributed seedlings ?
7. The lack of women's participation has not only effected the social forestry programme but also made life harder for rural families.

Suggestions

1. Grant of fresh mining leases to be discouraged specially in the vicinity of forest areas.
2. There is a need of undertaking employment and income generation programmes to reduce the dependence of tribals / villagers on forests .
3. The ratio of plant species providing Non Timber Forest Produce (NTFP) should be increased in the plantations and development of proper marketing facilities and infrastructure for processing of NTFP and value addition for better economic gains to the rural masses.

4. In order to save the limited forest resources of the district, non-conventional energy inputs e.g. solar cookers, and heaters and use of fuelwood saving divices should be encouraged.
5. The forestry extension is one of the aspects in which the forest department (F. D) in the district is not fully equipped. It is essential to set up independent and dedicated extension units at the district level and should have sufficient staff and equipments such as EDP, Audio-visual aids and vehicles.
6. Agro-forestry on private lands should be encouraged, since it is a combined production system. So it is a step in the right direction to evolve an integrated land use pattern.
7. Only indigenous and locally suitable tree species of multiple use should be raised, under the farm forestry programme.
8. Villagers have intimate knowledge and understanding of the local resources *e.g.* forests, water, and tree species etc. Therefore due respect and use of people's knowledge in planting should be aclaimed.
9. Foresters have to pay more attention to the expectations and needs of the people and involvement of local population in managing the forest sustainability.

Looking to the above explanations and the analytical study, it may be said that it is (S. F.) a scheme for the people, by the people and for the people to protect environment and the future. But it needs all seriousness, adequate finance, and popularity for its real success.

Appendix I

List of Common Plants Found in the Bhilwara - District

a. Trees

S. No.	*Local name*	*Botanical name*
1.	Aal	Morinda tinctoria, Roxb.
2.	Aam, Amba	Mangifera indica, Linn.
3.	Achar, Chiraunji	Buchanania lanzan, spreng.
4.	Akol	Alangium salvifolium (Linn. f.), Wang.
5.	Amaltas, Karmalia	Cassia fistula, Linn.
6.	Anjan	Hardwickia binata, Roxb.
7.	Aonla	Emblica officinalis, Gaertn.
8.	Aranda	Ricinus Communis, Linn.
9.	Ardu	Ailanthus excelsa, Roxb.
10.	Arithi	Sapindus emarginata, Vahl.
11.	Ashapala, Asok	Polyalthia longifolia, Benth & Hk.
12.	Babul, Bawal	Acasia arabica, Wild.
13.	Bad	Ficus bengalensis, Linn
14.	Bahera, Beda	Terminalia belerica, Roxb.
15.	Bakain	Melia azadirach, Linn.
16.	Bans	Dendrocalamus strictus, Nees.
17.	Barna	Cartaeva religiosa, Forst.

S. No.	Local name	Botanical name
18.	Bija, Biya	Pterocarpus marsupium, Roxb.
19.	Billi, Bel	Aegle marmelos, Corr.
20.	Bistendu, Kara-timru	Diospyros montana.
21.	Bor	Zizyphus Jujuba Lamk.
22.	Bukhan	Parkinsonia aculeata, Linn.
23.	Candan	Santalum album, Linn.
24.	Churel	Holoptelea integrifolia, Planch.
25.	Dhak, Khankra	Butea monosperma, (Lamk.)
26.	Dhaman	Grewia Vesta, Wall.
27.	Dhaura	Anogeissus latifolia, Wall.
28.	Dhokra, Kaldhi	Anogeissus pendula, Edgew.
29.	Dudhi, Karontya, Karu	Holarrhena antidysenterica, Wall
30.	Farash	Tamarix dioica Roxb.
31.	Gadapalash	Erythrina indica, Roxb.
32.	Gadi hagro	Kydia calycina, Roxb.
33.	Ghat bor	Zizyphus glaberrima, Santapau.
34.	Garnal	Cochospermum religiosum, (Linn.) Alston.
35.	Godal	Lannea coromandelica, (Houtt.) Merr.
36.	Goya Khair	Dichrostachys cinerea, W. & A.
37.	Gugal	Commiphora mukul (Hook. ex Stocks.) Engl.
38.	Gular, Umar	Ficus glomerrata, Roxb.
39.	Gundi (Bar), Lasora	Cordia Dichotoma, Forst.
40.	Gundi (Rai), Lasora	Cordia rothii, R. & S.
41.	Haldu	Adina cordifolia, Hook, f.
42.	Hingotia	Balanites aegyptiaca, (Linn.) Delile.
43.	Imli, Aamli	Tarmarindus indica, Linn.

S. No.	*Local name*	*Botanical name*
44.	Jal, Pilu	Salvadora oleoides, Dcne.
45.	Jal, Khara	Salvadora persica, Linn.
46.	Jamnai	Salvadora jambos, (Linn.) Alston.
47.	Jamun, Jambua	Syzygium cumini, (Linn.) Sleels.
48.	Jangal jalebi	Inga dulcis, Willd.
49.	Jhinja	Bauhinia racemosa, Lam.
50.	Kachnar	Bauhinia Variegata, Linn.
51.	Kainth, Kabeeta	Feronia limonia, (Linn.) Swingle.
52.	Kakon	Flacourtia indica (Burm. f.) Merr.
53.	Karanja, Kanjeri	Pongamia pinnata, (Linn.) Pierre.
54.	Kalam, Kamra	Mitragyna Parvifolia (Roxb.) Korth.
55.	Karaya, Kara	Sterculia Urens, Roxb.
56.	Ker	Capparis decidua, (Forsk.) Pax.
57.	Kela	Musa sapientum, Linn.
58.	Kevra	Pandanus Fascicularis, Lam.
59.	Khair	Acacia catechu, Wild.
60.	Khajur	Phoenix sylvestris, Roxb.
61.	Khirni, Dudhi	Wrigthtia tomentosa, Roem. & Sch.
62.	Khejri.	Prosopis spigeria, Linn.
63.	Kumta.	Acacia senegal, Wild.
64.	Lankpana, Kaljharia	Bridelia retusa, Spreng.
65.	Mahua, Mahura	Madhuca indica, Gmel.
66.	Mendala	Randia dumetorum, Lamk.
67.	Mokha	Schrebera, swietenioides, Roxb.
68.	Neem, Limra	Azadirachta indica, Juss.
69.	Nimboo	Citrus media, Linn.
70.	Padal	Stereospermum snaveolens, DC.
71.	Pader	Gargenia turgida, Roxb.
72.	Pai	Dalbergia lanceolaria, Linn.

S. No.	*Local name*	*Botanical name*
73.	Paras pipal	Ficus cordifolia, Roxb.
74.	Pakhar	Ficus lacor, Buch-Ham.
75.	Pipal	Ficus religiosa, Linn.
76.	Ratanjot	Clausena pentaphylla, DC.
77.	Raen (Khirni)	Mimusops hexandra, Roxb.
78.	Renjua, Khejra	Acacia leucophloea, Wild.
79.	Rohan, Royan	Soymida febrifuga, Juss.
80.	Rohani, Rolli	Mallotus phlippinesis, muell.
81.	Rohera	Tercomella undulata, (Smith.), Seem
82.	Sadar	Teminalia tomentosa, W. & A.
83.	Sag, Sagwan	Tectona grandis, Linn. f.
84.	Salar, Halar	Boswellia serrata, Roxb.
85.	Sandeshra	Delonix elata, (Linn.) Gamble.
86.	Sargora	Moringa concanensis, Nimm.
87.	Semal, Hemlia	Salmalia malabarica, (DC) Schott & Endl.
88.	Senjna, Surjna	Moringa oleifera, Lamk.
89.	Sewan, Havan	Gmelina arboria, Linn.
90.	Shisham	Dalbergia Sissoo, Roxb.
91.	Shisham (Kala)	Dalbergia latifolia, Roxb,
92.	Shahtoot	Morus alba, Linn.
93.	Siris (Safed), Garar	Albizzia procera, Benth.
94.	Siris (Kala), Harsya	Albizzia lebbek, Benth.
95.	Sitaphal	Anona Squamosa, Linn.
96.	Tambolia	Ehretia laevis, Roxb.
97.	Timru	Diospyros melanoxylon, Roxb.
98.	Tanaj	Ougeinia dalbergioides, Benth.
99.	Umbia	Saccopetalum tomentosum, Hk. f & T.
100.	Vanakrya	Garuga pinnata, Roxb.
101.	Vilayati Babul	Prosopis juliflora, Hk. f. & T.

b. Shrubs and Herbs

S. No.	*Local name*	*Botanical name*
102.	Adhashishi	Xanthium Strumarium, Linn
103.	Aak	Calotropis procera, Br.
104.	Aakra	Calotropis gigantea, Br.
105.	Andhijhara	Achyranthes aspera, Linn.
106.	Anwal	Cassia auriculata, Linn.
107.	Arakpush	Gynadropsis gynandra (Linn.) Briq.
108.	Arni	Clerodenron viscosum, Vent.
109.	Arusa	Adatoda vasica, Nees.
110.	Baibarrang	Embelia tsjeriam-cotton (R. & S.) A. DC.
111.	Bajardanti	Barleria Prionitis, Linn.
112.	Banharar	Flemingia semialata, Roxb.
113.	Bankapas	Thespesia lampas, Dalz. & Gibs.
114.	Bapsi	Ocimum americanum, Linn.
115.	Basuni	Tephrosia hookeriana, W. & A.
116.	Bekal	Gymnosporia Spinosa, (Forsk.) Fiori.
117.	Bekari	Indigofera cordifolia, Heyne.
118.	Bakari	Indigofera linifolia, Retz.
119.	Bekario	Indigofera anabaptista, Steud.
120.	Bhang	Cannabis sativa, Linn.
121.	Bhurangini	Solanum xanthocarpum, Schard & Wendi.
122.	Bichhu	Martynia annua, Linn.
123.	Bui	Aerua tomentosa, Forsk.
124.	Chirion-ka-dhaniya	Mollugo Cerviana, Ser.
125.	Damasa	Fagonia cretica, Linn.
126.	Dasan	Rhus mysorensis, Heyne.
127.	Dhatura	Datura metel, Linn.

S. No.	*Local name*	*Botanical name*
128.	Dhawri	Woodfordia fruticosa, (Linn.) Kurz.
129.	Dholekan	Grewia villosa, Wild.
130.	Dudhi	Euphorbia hirta, Linn.
131.	Gangan	Grewia tenax, (Forsk) Fiori.
132.	Gursharia	Grewia hirsuta, Vahl.
133.	Gokhru	Tribulus hirsuta, Linn.
134.	Indrayan	Citrullus colocynthis, Schrad.
135.	Jawasa	Alhagi camelorum.
136.	Jharber, Bor	Zizyphus nummularia, (Burm. F.) W. & A.
137.	Jhau	Tamarix dioica, Roxb.
138.	Jhil	Indigofera oblongifolia, Forsk.
139.	Jhinjani	Mimosa rubicaulis, Lam.
140.	Kali siali	Grewia flavescens, Juss.
141.	Ḳalwari, Jal	Capparis spinosa, Lamk.
142.	Kanter	Capparis sepiaria, Linn.
143.	Karonda	Carissa spinarum, A. DC.
144.	Khimp	Periploca aphylla, Dcne.
145.	Kherata	Dodonea viscosa, Linn.
146.	Khimpura	Leptadenia pyrotechnica, (Forsk.) Dcne.
147.	Makoy	Solanum Nigrum, Linn.
148.	Makoh	Zizyphus oenoplia, Mill.
149.	Mendi	Lawsonia Inermis, Linn.
150.	Marorphali	Helicteres isora, Linn.
151.	Murali	Lycium europaeum, Linn.
152.	Musli (Safed)	Chlorophytum tuberosum, Roxb.
153.	Nag-phanithor	Opuntia dillenii, Haw.
154.	Neel	Indegofera tinctoria, Linn.
155.	Negar, Nirgundi	Vitex negundo, Linn.
156.	Pawaar	Cassia tora, Linn.

S. No.	*Local name*	*Botanical name*
157.	Pithari	Abutilon ramosum, Guill. & Perr.
158.	Salpan	Securinega virosa, (Roxb. ex Willd) Pax & Hoffm.
159.	Satialeti	Glinus lotoides, Linn.
160.	Satyanasi	Argemone mexicana, Linn.
161.	Senia	Crotalaria burhia, Ham.
162.	Tamat, Harsingar	Nyctanthes arbortristis, Linn.
163.	Thor	Euphorbia Nivulia, Ham.
164.	Thor	Euphorbia royleana, Boiss.
165.	Tijara	Papaver somniferum, Linn.
166.	Tulsi	Ocimum Sanctum. Linn
167.	Tilbam	Wendlaudia exserta, DC.
168.	Unt kantalo	Lepidagathis trinervis, Nees.
169.	Tulsi, Bantulsi	Ageratum conyzoides, Linn.

c. Climbers

170.	Amarbel	Cuscuta reflexa, Roxb.
171.	Bajarbel	Cocculus hirsutus, (Linn.) Diels.
172.	Chirmu, Ratti	Abrus precatorius, Linn.
173.	Dudhi, Kalibel	Ichnocarpus futescens, Br.
174.	Dudhi	Vallaris solanocea, (Roth) O. Ktze.
175.	Dujni	Millettia auriculata, Baker.
176.	Ghorabel	Pueraria. Spp.
177.	Gurbel	Viscum orentale, Willd.
178.	Hirkiabelia	Acacia pennata, Willd
179.	Hukhabel	Aristolochia bracteata, Retz.
180.	Kangni, Malkangni	Celastrus paniculata, Willd.
181.	Katamba	Cayratia carnosa, Gagnep.
182.	Kenwach	Mucuna prurita, Hook.
183.	Maljhan	Bauhinia vahlii, W. & A.
184.	Medhasingi, Nagbel	Cryptolepis buchanani, R. & S.

S. No.	*Local name*	*Botanical name*
185.	Nak chinkni	Clematis Roylei, Rehder.
186.	Narkanta	Asparagus Dumosus, Baker.
187.	Nahbel	Butea parviflora, Roxb.
188.	Neem giloy	Tinospora Cordifolia, Miers.
189.	Paharbel	Cissampelos pareira, Linn.
190.	Plasbel	Butea Superba, Roxb.
191.	Pilwaribel	Cocculus pendulus, (Forst.) Diels.
192.	Tripatti	Oxalis corniculata, Linn.

d. Grasses

193.	Anjan, Dhaman	Cenchrus ciliaris, Linn.
194.	Bajra	Pennisetum typhoides, (Burm. f.) Stapf & C.E. Hubb.
195.	Bamna	Cloris dolichostachya, Lag.
196.	Baru	Sorghum halepense, (Linn.) Pers.
197.	Bharut	Cenchrus setigerus, Vahl.
198.	Bhangta, Bhero, Polad	Apluda mutica, Linn.
199.	Chiri-Bajra	Eragrostis tremula, Hochest.
200.	Dab	Imperata cylindrica, (Linn.) Beauv.
201.	Dub	Cynodon dactylon, Pers.
202.	Gundar	Cymbopogon parkeri, stapf.
203.	Kaiyo	Coix-gigantea, Koenex Roxb.
204.	Kans	Saccharum spontaneum, Linn.
205.	Khas	Vetiveria Zizanioides, (Linn.) Nash.
206.	Lampla	Aristida depressa, Retz.
207.	Munj	Saccharum munja, Roxb.
208.	Nal	Phragmites karka, (Retz.) Trin, ex Steud.
209.	Ratarra, Bheri	Themida quardrivalvis, O. Ktze.
210.	Rohira, Rosha	Cymbopogon martinii, (Roxb.) Wats.

S. No.	Local name	Botanical name
211.	Sawan	Echinochola colonum, (Linn.) Link.
212.	Seran	Chrysopogon fulvus, (Spreng.) Chiov.
213.	Sien	Sehima nervosum, (Rottl.) Stapf.
214.	Sitaghas	Tragus biflorus, (Roxb.) Schult.
215.	Soorwala	Heteropogon contortus, (Linn.) Beauvex R. & S.
216.	Jharnia, Makra	Dactyloctenium Aegyptium, (Desf.) Beauv.
217.	Karar	Dichanthium annulatum, (Forsk) Stapf.
218.	Jungli	Eremopogon Foveolatus, (Del.) Stapf.
e. Parasites		
219.	Banda	Dendrophthoe falcata, (Linn. f. Etting. Syn. Loranthus Longiflorus Desr.

Source : Working Plan for the Chittorgarh (Bhilwara), Forest Division, 1962.

Appendix II

List of Common Birds Found in the Bhilwara - District

S. No.	English name	Zoological name
1.	Little Grebe	Podicaps ruficallis.
2.	Little Cormorant	Phalacrocorax niger.
3.	Indian Darter	Ankinga rufa.
4.	Grey Heron	Aredes cigerea.
5.	Pond Heron, Paddy bird	Ardeola grayil.
6.	Cattle Egret	Buoulcus ibis.
7.	Little Egret	Egretta qurzetta.
8.	Painted Stork	Ibis leucocephalus.
9.	Whitenecked Stork	Ciconia eiscopus.
10.	White Ibis	Threskiornis melanocephola.
11.	Spoonbill	Platales leucorodia.
12.	Lessor Whistling treal	Dendrocygna javenica.
13.	Rubby Sheldrake	Tadorna ferryginea.
14.	Paintail	Anas scuta.
15.	Spotbill	A. Poeoilorhyncha.
16.	Cotton Teal	A. Crecca.
17.	Nukta	Sarkindiornis melanotes.
18.	Blackwinged Lits	Elanus caeruleus.
19.	Crested Honey Buzzard	Pernis ptilorbynchus.

S. No.	*English name*	*Zoological name*
20.	Pariah Kite	Milyus migrans.
21.	Shikara Kite	Accipiter badius.
22.	White-eyed Buzzard	Butastur teesa
23.	King Vulture	Tergus Calvus.
24.	White Backed Vulture	Gyps bengalensis.
25.	Tawny eagle	Aquila repax.
26.	White Scavanger Vulture	Neophron perenopterus.
27.	Kestrel	Falce Tinnunculus.
28.	Black Partridge	Francolinus francolinus.
29.	Gray Partridge	F. Pondicerianus.
30.	Rain Quail	Coturnix coromandelica.
31.	Jungle Bush Quail	Predicula asiatics.
32.	India Peafoul	Paw cristatus.
33.	Sarus Crane	Grus antigone.
34.	White Breasted Waterhen	Amaursris phoenicurus.
35.	Moorhen	Gallinula Chloropus.
36.	Purple Moorhen	Porphyrio porphyrie.
37.	Common Coot	Fulica atra.
38.	Pheasant-Tailed Jacana	Hydrophasienus chirurgus.
39.	Red-wattled Lapwing	Vanellus indicus.
40.	Yellow-wattled Japwing	V. malabaricus.
41.	Red Shank	Tringa totanus.
42.	Wood sand piper	Tringa Glareola.
43.	Common sound piper	T. hypolecucos.
44.	Little Stint	Calidris munuta.
45.	Black wing stilt	Himantophus himantopus.
46.	Curlew	Numenius arquata.
47.	Stone curlew	Burhinus oedicnemus.
48.	Indian courser	Cursorius coromandelicus.
49.	River Tern	Sterna aurantia.
50.	Common sandgrouse	Pteroclas exustus.

S. No.	English name	Zoological name
51.	Green pigeon	Treron phoenicoptera.
52.	Blue Rock Pigeon	Columba olivia.
53.	Red Turtle Dave	Streptopelia tranqueberic.
54.	Indian Ring Dove	Streptopelia decaocta.
55.	Spotted Dove	Streptopelia chinensis.
56.	Little Brown Dove	Streptopelia bengalensis.
57.	Large Indian Parakeet	Psittacula eupatoria.
58.	Redringed Parakeet	Psittacula krameri.
59.	Blossom Headed Parakeet	Psittacula cyanocephal.
60.	Common Hawk Cuckoo	Cuculus varius.
61.	Pied crested cuckee	Clamator jacobinus.
62.	Koel	Eudynamys scalepacea.
63.	Coucal, Crowphyeasant	Centropus sinensis.
64.	Spottedowlst	Athen brama.
65.	Collard scops owl	Otus bakkamoena.
66.	Common Indian Nightjar	Caprimilous asiatic.
67.	Franklines Nightjar	C. affirmis.
68.	House Swift	Apus affinis
69.	Palm Swift	Cypsiurus parvus.
70.	Piid Kingfisher	Ceryle rudis.
71.	Common Kingfisher	Alcedo atthis.
72.	Whitebreasted Kingfisher	Haleyon smyrensis.
73.	Billed Kingfisher	Pelargopsis capensis.
74.	Common green bee eater	Merops orjentalis.
75.	Bluetailed bee eater	Merops philippinus.
76.	Chesthut-headed	Merops leschenault.
77.	Bee-eater	M. apiaster.
78.	Indian Roller	Coracias benghalen.
79.	Suropeon Roller	C. garrulus.
80.	Hoopoe	Upupa epops.
81.	Grey Hornbill	Tockus birostris.

S. No.	*English name*	*Zoological name*
82.	Coppersmith	Megalamia heamacephala.
83.	Wryneck	Jynx torquilla.
84.	Goldenbacked wood pecker	Dinopium benghalensa.
85.	Yellow fronted pied woodpecker	Picoides mahrattensis.
86.	Indian Pitta	Pitta brachyura.
87.	Redwinged Bush lack	Mirafra erythroptera.
88.	Ashy Crowned Finch Lark	Ermoptsrise grisea.
89.	Rufous taqiled finch Lark	Ammomances phoenicuries.
90.	Creasted Lark	Galerida cristata.
91.	Dusky Crag Martin	Hirunde concolor.
92.	Wiretailed swallow	H. smithil.
93.	Red rumpted swallow	H. daurica.
94.	Grey shrike	Lanius excubitor.
95.	Bay backed shrike	L. vittatus.
96.	Rufousbacked shrike	Lanius schach.
97.	Golden oriole	Oriolus oriolus.
98.	Black Drongo, King crow	Dicrurus admimilbs.
99.	White beilied Drongo	D. caerulescens.
100.	Brahmni or Black headed Myna	Strurnus pagodarum.
101.	Rody Pasteor	S. roseus.
102.	Common Myna	Acridotheres tristis.
103.	Bank Myna	A. ginginianus.
104.	Indian Tree Pie	Dendrocitta vagabunda.
105.	House Crow	Corvus splandens.
106.	Jungle Crow	C. mecrohynchos.
107.	Black headed Cuckoo shrike	Coracina melaneptera.
108.	Large cuckoo shrike	C. novaehollandise.
109.	Wood Shrike	Tephrodensis pondicerianus.
110.	Scarlet Minivet	Pericrocotus flammeus.
111.	Small Minivet	P. cinnamomeus.

S. No.	English name	Zoological name
112.	Iora	Agithina tiphia.
113.	White Checked bulbul	Pycnoetus leucoqenys.
114.	Redvented bulbul	P. cafer.
115.	Yelloweyed Bulbler	Chtysomma sinense.
116.	Common bubbler	Turdoides caudatus.
117.	Large grey bubbler	T. malcolmi.
118.	Red breasted Flycatcher White brow	Miscicapa parva.
119.	Grey headed flycatcher	Gulicicapa ceyoloensis
120.	White browed fantail Flycatcher	Rhipidura aureola.
121.	Paradise Flycatcher	Terpsiphone paradisi.
122.	Franklin's ween-wrabler	Prinia Bodgsonil.
123.	Tailer Bird	Orthotomus sutorius.
124.	Lesser whitethroat	Sylyia curruca.
125.	Magpie Robin	·Copsychus saularis.
126.	Indian Robin	Saxicoloides fulicata.
127.	Black redslait	Phooenicurus ochruros.
128.	Browhn Rockhat	Cercomela Fusca.
129.	Collered Bushchat	Saxicola torquata.
130.	Pied Bushchat	S. caprata.
131.	Gray tit Yellow checke tit	Parus Major I. xanthrogenys.
132.	Yellowheaded wagtail	Motacilla citreola.
133.	Grey Wagtail	Motacilla caspica.
134.	White Wagtail	M. alba.
135.	Large Pied Wagtail	M. maderras patensis.
136.	Pruple Sunbird	Nectarinia asistica.
137.	White Eye	Aosterops palpebrosa.
138.	House Sparror	Passer domesticus.
139.	Weaver Bird Red	Ploceus phillppinus.
140.	White Throated Munia	Lonehura malabarica.
141.	Crosted Bunting	Helophus lathani.

Source : Working Plan for Bhilwara Forest Division, 1996.

Appendix III

List of Common Mammals Found in the Bhilwara - District

S. No.	*English name*	*Latin name*
1.	Rhesus Macaque, Monkey	Macaca mulatta.
2.	Common Langur Hanuman monkey	Presbytis entellus.
3.	Baghera	Pantnera pardus.
4.	Jungle cat	Felis chaus.
5.	Common Mongoose	Nerpestes edwerdsi.
6.	Jackal	Canis aureus.
7.	Indian Fox	Vulpes bengalensis.
8.	Striped Hyaena	Hyaena hyaena.
9.	Common Palm Civet	Paradoxarus hermaphooditus.
10.	Smooth Indian Otter	Lutra perspecillata.
11.	Blue bull.	Boselaphus tragocamelus.
12.	Indian Wild Beer	Susseroja cristatus.
13.	Five stribed Palm Squirrel	Funembulus pennanti.
14.	Common House Rat Indian Field Mouse.	Rattus rattus Mus booduga.
15.	Indian Porcupine	Hystrix Indica.

S. No.	*English name*	*Latin name*
16.	Indian Hare	Lepus nigricollis.
17.	Grey Musk Shrew	Sunous murinus.
18.	Bat	Peteropus giganteus.
19.	Four horned antelope	Tetracerus quadrieornis.
20.	Wolf	Canisa laupus.

Source : Working Plan for Bhilwara Forest Division, 1996.

Bibliography

Agarwal, Anil & Sunita Narain. 1985, The State of India's Environment. The second citizen's report.

Agarwal, S.K. 1986, Ecology of Chambal river at Kota. Acta Ecologia 8(1).

Amalara, J. and V.A. Shankarnarain. 1986, Ecological distribution of Balanites roxburghi. PI in arid Rajasthan. Jou. Tropical forestry, 2(3).

Anderson, M.S. 1971, The Geography of living things. University's Press London.

Anonymous. 1952, Government of India. National Forest Policy of India.

Anonymous. 1970, Govt. of India. Wild life conservation in India.

Anonymous. 1973, Govt. of India. National Commission on Agriculture's interim report on Social Forestry.

Anonymous. 1976, Fifteen years of Arid Zone Research. CAZRI Jodhpur.

Anonymous. 1978, Proceedings of the Farm Forestry Symposium, ICAR, New Delhi.

Anonymous. 1979, Tropical Legumes Resources for the Future. Nat. Acad. Sci. Washington, D. C.

Anonymous. 1981, Govt. of India. Development of Forestry and Forest Products. Community Profile, Ministry of Agriculture, New Delhi Govt. of India. Report of the fuelwood study committee planning commission, New Delhi.

Anonymous. 1982, Govt. of India. Report of the fuelwood study committee. Planning commission, New Delhi.

Anonymous. 1982, Govt. of India. Forestry in India Ministry of Agriculture, New Delhi.

Anonymous. 1983, A bridged Glossary of Technical Terms. F.R.I. Dehradun.

Anonymous. 1984, Govt. of India. Report of the committee of Rights and Concessions in Forest Areas of India.

Archer, B. *et. al.* 1976. Understanding Biology. Mills and Boon; London.

Arvil. R. 1976, Man and Environment. A pelican original Middledsex, U.K.

Audus, L.J. 1953, Plant Growth Substances. Leonard Hill, London.

Bachketi, N.D. 1984, Social Forestry in India, Problem and prospects. Published by Birla Institute of Scientific Research, New Delhi.

Backer, F.S. 1950, Principles of Silviculture. MC. Graw Hill Book Co., New York.

Bakshi, B. K. 1984, Forest Pathology, Principles and Practices in Forestry. Controller of Publications Delhi.

Banks, H.P. 1970, Evolution and Plants of past. Macmillan, London.

Bender, B. 1975, Farming in Pre-history. John Baker London.

Bhatia, G.N., Kala, J.C., and Vyas, D.K. 1986, Economic appraisals of date palm plantations in western Rajasthan. Ann. Arid Zone, 24(5).

Bhat, L.S. 1976, Micro-level planning. A case study of Karnal and New Delhi.

Bhandari, M.M. 1978, Flora of the Indian Desert. Scientific Publishers, Jodhpur.

Bhola, P.P. and Sharma, Y. K. 1982, Kraft pulping of Leucaena leucocephala. Indian Forester 108(3), pp. 202-211.

Billings, W.D. 1964, Plants and Ecosystem. Macmillan, London.

Bose, R.K. and Bandopadhyay. 1986. Economics of energy plantations in alkali soils of Indian semi arid regions. Biomass 11(1).

Brady, N. C. 1965, The Nature and Properties of soils. Macmillan; New York,

Bunting, B. T. 1965, The Geography of Soil. Hutchinson, London.

Champion & Seth. 1968, General Silviculture for India. Govt. of India Publication, Delhi.

Champion & Seth. 1968, A Revised Survey of Forest Types of India.

Chaturvedi., A.N. 1985, Firewood Farming on Degraded Lands. U.P. Forest Bulletin No. 50.

Chaturvedi, M.P. 1946, Land Management in United Provinces. Supt. Printing & Stationery U.P.

Chatterji., P. C. 1985, Impact of Human Activities on the Water Resources of the Arid Zone. Prog. Develop. Management Training Course on Irrigated Agriculture in Arid Areas, held at CAZRI, Jodhpur, pp. 135-141.

Chauhan, B.J. 1966, Studies in Land Utilization.

Clements, F. E. 1939, Bio-ecology. Mc Graw Hills, New York.

Cloudsley, Thompson, J.L. 1977, Man and Biology of Aridzones, Edward, Arnold London.

Cole, M. M. 1975, Recent Development in Geography. Longman, London.

Darlington, P. 1957, Zoogeography., Methuen, New York.

Das Gupta, B. 1971, Socio-Economic Classification of Districts. A Statistical Approach. Economic and Political Weekly.

Dav, Hullah and Kaul, V. 1986, Agroforestry, A possible way out of the increasing human predicament. Jou. Trop. Forestry 2(3).

Dhabriya, S.S. 1981, Maping and Analysis of Resources by Satellite imaginary III Indian Geo. Congress J.N.U. Delhi.

Dhabariya, S.S. 1982, Delineation of Geomorphic Regions and Prospects for Development in Bhilwara District.

Elton, C.S. 1958, The Ecology of Invasion of Plants and Animals. Methuen, London.

Etherington, J.R. 1975, Environment & Plant Ecology. John Wiley, London.

Etherington, J.R. 1978, Plant Physiological Ecology. Arnold, London.

Evans, D. P. and Thomas, J. L. 1981, Water in Desert Ecosystem. Dowden Hutchingon & Ross, Inc. Stroudsburg.

Falkner, F.R 1984, The Desert Limit of Dry Farming in Africa. Geographer.

Fogg, G.E. 1970, The Growth of Plants. Penguin Books, Harmondsworth.

Foth, H.O. and Turk, L. M. 1972, Fundamentals of soil science, Wiley, New York.

F. S. I. 1993, The state of Forest Report. F. S. I., Govt. of India, New Delhi.

Garnier, B. J. 1976, Methods and Perspective in Geography. Longman London.

Ghosh, R.C. 1977, Hand book of Afforestation Techniques. F.R.I. Dehradun.

Gohil, R.N. 1985, Recent Trends in Botanical Research.

Golley, F. B., and Medina, F. 1975, Tropical Ecological Systems. Ecological studies 11, Springerverlag, New York.

Good, R.D. 1953, The Geography of Flowering Plants. Longman London.

Goudie, A. 1990, The Human Impact on the Natural Environment. Oxford London.

Govt. of India. 1988, The National Forest Policy. Govt. Press, New Delhi.

Graham, S.A. 1952, Principles of Forest Entomology (3rd. ed.) Mc. Graw Hill Book Co., New York.

Grigg., D.B. 1974, The Agricultural Systems of the World. An Evolutionary Approach, Cambridge University Press, London.

Gupta, P.D. 1983, Bio-chemical studies on plant Tissue Culture. Studies with Eu. citriodora, Ph. D. Thesis, Poona Univ. (p).

Iyer, Justice, V.R. Krishna. 1984, Environmental Pollution and the Law.

Joshi, H. B. 1980, The Silviculture of Indian Trees. F. R. I. Dehradun.

Joy, T.V. 1982, Biogeography, Longman London.

Kathiresan, S. Bala. 1986, Essentials of Forest Management. Natraj Publishers Dehradun.

Konds, S. 1981, Criteria for mass selection of Fuelwood species with reference to Agricultural resources of Tamil Nadu.

Kotia, K.K. 1962, Working plan for the Chittorgarh (Bhilwara) Forest Division.

Kozlowski, T.T. 1968, Water deficits and Plant Growth. Mac Graw, Hill, New York.

Lal, A.B. 1967, Indian Silviculture. Jugal Kishore & Co. Dehradun.

Levit, J. 1972, Responses of Plants to Environmental stresses. Academic press New York.

Mac Arthur, R.H. 1984, Geographical Ecological Patterns in the distribution of species. Princeton Univ. Press New Jersy.

Maheshwari, J.K. 1976, Plant Resources of India. Sci & Cult 42.

Maithani, G.P. and Bahuguna V.K.P. Lal. 1986, Effect of Forest fires on the ground vegetation of a moist deciduous Sal Forest. Ind. Forester 112 (8)

Maslekar, A.K. 1981, Forester's Companion. Jugal Kishore & Co. Dehradun.

Maslekar, A.R. 1983, Managing the forests, Jugal Kishore & Co. Dehradun.

Meher-Homji, V.M. 1986, Disasters of Deforestation. Int. J. Eco. Environ, Sc. 10.

Mishra, V.C. 1967, Geography of Rajasthan. N.B.T., New Delhi.

Muller, P. 1986, Biogeography. Harper and Row, New York.

Oosting, H. J. 1956, The Study of Plant Communities. W. H. Freeman, San Francisco.

Pathak, P.S., and Patil, B.D. 1986, Grow stylo on Waste Lands. Indian Farming, 35(12).

Pears, N. 1985, Basic Biogeography, Longman London.

Polunin, N. 1960, Introduction to Plant Geography. Mc Graw Hill, New York.

Puri, G. S. 1960, Indian Forest Ecology. Oxford Book & Stationery Co., New Delhi.

Raunkiaer, C. 1934, The Life Forms of Plants and Statistical Plant Geography. Clarendon Press, Oxford, London.

Raza, Moonis. 1981, Man in the Eco-system, The Geographical View. Occasional Paper C.S.R., D., JNV P. 5.

Robinson, G.W. 1951, Soil, Their Orian, Constitution and Classification. Longman London.

Robinson, H. 1982, Biogeography, Macdonald and Evans London.

Russel, E.W. 1973, Soil conditions and plant Growth. Longman London.

Sakerwal, K.C. 1996, Working Plan for the Bhilwara Forest Division.

Sauer, C.O. 1969, Agricultural Origion and Dispersal. M.I.T. Press, Cambridge.

Saxena, V.S. 1993, Tree Nurseries and Planting Practices, Delta Printing Bombay.

Saxena, V.S. 1993, Useful Trees, Shrubs and Grasses. Delta Printing Bombay.

Seddon, B. 1971, Introduction to Biogeography. Duck Worth London.

Shah. S.A. 1988, Forestry for people, I.C.A.R. New Delhi.

Sharma, B.L. 1983, Agricultural Typology of Rajasthan. Udaipur, (Raj.)

Sharma, B.L. 1991, Dry Land Farming, Daya Publication, New Delhi.

Sharma, B.L. 1994, Nature of Geography as Perceived in Systems Theory. Daya Pub., New Delhi.

Sharma, Y.M.L. 1959, Lessons in Forestry. I.C.A.R., New Delhi.

Shukla, N.K. & Rajput, S. S. 1983, Physical and Mechanical properties of Eucalyptus grown in India. Indian Forester 109 (12) pp. 933-943.

Solanki, M.S. 1981, Forests as a source of food, F.A.O. of U.N.O. Regional office for Asia and Pacific. Bangkok.

Srinath, R.A. and Kulkarni, H.D. 1986, Sandal in Dry Zone Afforestation. Geobios, 13 (203).

Tiwari, K.M. 1983, Social Forestry in India. Natraj Publishers, Dehradun.

Tiwari, R.C. and Tripathi, S. 1987, Concept of Environment in ancient India. International symposium, on Geomorphology and En. Management, Allahabad.

Vishnoi, R.L. 1987, "Samajik Vaniki Se Samrdhi". (Hindi), Bhilwara.

Vyas, N.N. (ed). 1981, Social Forestry in Tribal Development, M.L.V. Tribal Research and Training Institute, Udaipur.

Ward, H.M. 1980, Grasses, Scientific Publishers, Jodhpur.

Yadav, J.S.P. 1960, Soils of the Dry Zone of India. Ind. Forester 86(5).

Subject Index

Author Index